WARRIOR WITHDRAWAL

WARRIOR
WITHDRAWAL

When BAMF No Longer Means
Bad*$$ M^ther#u@!er

VARPAS DE SA PEREIRA, PsyD

Ballast Books, LLC
www.ballastbooks.com

ISBN: 978-1-962202-50-3

Printed in the United States of America

Published by Ballast Books
www.ballastbooks.com

For more information, bulk orders, appearances, or speaking requests, please email: info@ballastbooks.com

To those who were not identified and were lost to the syndrome and to the families that miss their loved ones because of it.

Then I heard the voice of the Lord saying: "Whom shall I send? Who will go for us?" and I said: "Here am I. Send me!"

—Isaiah 6:8

He who has a why can bear almost any how.

—Nietzsche

TABLE OF CONTENTS

Introduction   i

PART I: Combat Hunger   **1**

Chapter 1 – Preparing for Combat   3

Chapter 2 – On the Way to War   11

Chapter 3 – Entering Enemy Territory   25

Chapter 4 – Death Stares Me in the Face   31

Chapter 5 – More Combat   47

Chapter 6 – Medals   61

PART II: War and Family   **73**

Chapter 7 – Dating through Emails   75

Chapter 8 – Family Life and Deployment   85

Chapter 9 – On the Road to Retirement   101

PART III: Identity Crisis   **111**

Chapter 10 – Hanging up the Uniform   113

Chapter 11 – Maslow's Hierarchy   127

PART IV: Combat Addiction   **135**

Chapter 12 – Warrior Withdrawal Syndrome: Constellation
        of Symptoms   137

Chapter 13 – Seeking Machismo   153

Chapter 14 – Seeking Fame   159

Chapter 15 – Seeking Tribe   163

Chapter 16 – Seeking High   169

Chapter 17 – Seeking Purpose   173

PART V: Retraining   **175**

Chapter 18 – Becoming a Psychologist for Vets with WWS   177

Chapter 19 – Diagnosis and Treatment at the VA   187

Chapter 20 – Treating Veterans at the VA   195

Chapter 21 – Running WWS Groups   209

Epilogue   229

Acknowledgments   231

Notes   233

INTRODUCTION

Life is not primarily a quest for pleasure, as Freud believed, or a quest for power, as Alfred Adler taught, but a quest for meaning.

—Victor Frankl, *Man's Search for Meaning*

I am a clinical psychologist at the VA in Long Beach, California, treating mostly former military men. I spend my day, by and large, sitting in a comfy chair, listening attentively to troubled souls spill their guts as they relive the traumatic events of their deployments. Some are maimed physically; all are maimed psychologically.

Being a psychotherapist contrasts sharply with my former role as a Marine major, deployed in Iraq for over fifteen years. There, my days were passed in dry, unbearable 130-degree desert heat, with sand rushing by and hot air blowing on my face like a hair dryer on full blast. Weighted down with equipment and water bottles from which I drank endlessly to quench my never-ending thirst, I was on constant alert for the danger that lurked around every war-torn corner.

Yet, I loved being there. Deployment suited me, although, at the time, I didn't understand why. I returned home and went from wearing a gas mask and a coyote-tan, two-inch bulletproof vest in the stiflingly hot Iraqi desert while being shot at to being Mr. Mom and changing the diapers of my two-year-old baby girl. Unsurprisingly, I suffered "separation anxiety" from the military and found it hard to adjust to civilian life.

While wiping my daughter's butt, the 2008 Oscar-winning film *The Hurt Locker* came to mind. The main character, who is part of an explosive ordinance disposal team during the Iraq War, loves the danger he experiences defusing bombs, even those wrapped around suicide bombers. When he returns home to his ex-wife and infant son, he finds the tedium of civilian life unbearable. In the last scene, he's back in Iraq for another year-long tour of duty defusing bombs with Delta Company.

When I saw the movie in the theater, people walked out muttering, "Why did he return? Makes no sense." Others wondered, "Was that guy nuts? Must have been a masochist."

I had no such thoughts. To me, his return to danger in Iraq made perfect sense. That was where he felt most alive.

War energized me, too. I lived with my wonderful wife and fabulous children in a lovely two-story, three-bedroom home in Orange County, with a "California-size" backyard for my kids to play in, located in a city with magnificent beaches and green parks. But some part of me longed to return to that dangerous desert terrain and again stare death in the face. Without danger, threat, and excitement, my life felt like a lie, and I struggled to find purpose. While the uniform was stored in a closet, what was the point of it all? Who was I?

What was going on? Did other vets feel this way, or was it a quirk within me?

In my last months on active duty in 2014, I spoke with a Green Beret soldier also in the process of leaving the service. Like me, he wanted to continue to deploy, even though he had a devoted wife waiting for him at home. Leaving danger depressed him.

"I get it," I told him. "Right now, I'm preparing for interviews for management consulting after retirement while also trying to do paramilitary operations for the CIA or the DOD."

Later I learned this desire to return to combat was a common sentiment in those of us who served; many vets I worked with said they would return to Iraq or Afghanistan in a heartbeat. Recently, many U.S. veterans have raced to fight in Ukraine.

To the non-military, this desire seems bizarre. Even our spouses struggled to understand it. Why would any sane person wish to trade air-conditioning for sweltering in the dry desert heat; homemade meals for disgusting food inside a plastic container; comfortable clothing for filthy, sweaty cammies; relative safety for life-threatening danger?

The military experience, whether in combat or not, is, in common Marine parlance, a fucking bitch. Deployments can be long, even years, and you miss out on comfort, security, families, and breaks. I was deployed in Iraq for three years cumulatively, for six months at a time without furlough. I missed the parental joy of seeing momentous milestones. While sweating it out in the desert, my son was born. While my oldest daughter was learning to crawl and to walk, I was in the middle of the operator training course for joint special operations command (JSOC), bouncing around the U.S. on training missions. I was present for only two weeks of my second daughter's first year and a half of life.

There's no end to the list of difficulties of life in the military. The armed forces dictate control over your life—where you live, what you wear, your daily routine, and your military occupational specialty (MOS). You deal with the same people day after day, eating with them, working with them, living with them, swearing with them, joking around with them. No change, no breaks, no escape, no women—just the guys, the battalion. If someone annoys you, tough luck. Forget about retreating to your private home or a weekend getaway.

Why would anyone be dying (pun intended) to get back to war?

This question baffles civilians, but almost every Marine I knew wanted to deploy. One of my peers said to my wife, "Your husband's string of continuous combat deployments is what every Marine wishes for."

I don't know that this feeling ever goes away. Patients I've worked with from the Vietnam conflict describe this fire burning for over fifty years. Some try to extinguish it with alcohol and other addictive substances—even with suicide.

It was not that I had a death wish or didn't love my family. Quite the opposite. Rather, my "combat envy" was an unexplainable urge that, at that time, I had no words for. Something was "there" that I couldn't pinpoint. During my conversation with the Green Beret, I had struggled to come up with a description that would capture the magnetism of deployments and combat. All that made sense was that it felt like an addiction. He agreed. Still, intuitively, I felt there was more than combat addiction. But what?

Then I came across *Man's Search for Meaning* by Viktor Frankl, an Austrian Jew and psychiatrist who survived the concentration camps. I was struck by how, like the lives of the former inmates of concentration camps, life had lost all meaning for returning vets.

In 1942, Frankl and his family were sent to the Theresienstadt concentration camp. His father died there of starvation and pneumonia. In 1944, Frankl and the surviving members of his family were transported to Auschwitz, where his mother and brother were murdered in the gas chambers. His wife died later of typhus in Bergen-Belsen.

Frankl spent three years in four concentration camps. He survived because he felt he had an important task to complete: the completion of a manuscript he had been working on. That manuscript later became *Man's Search for Meaning*. The book explains his psychological theory, called "logotherapy," which holds that a fundamental basis for psychological and spiritual health is a sense of purpose.

A lightbulb went off. Lack of *raison d'être* was a huge missing piece in vets adjusting to civilian life. In the military, they had purpose: to serve their country and save lives. When they came home, they were largely irrelevant; home felt like a foreign land. They were "like a rolling stone, without a home, like a complete unknown," as Bob Dylan captures in his classic lyrics.

Assisting these vets to find that missing piece and to give them hope for carving a path to a brighter future is what would give *my* life meaning, along with helping their spouses and children understand them better.

After much research and working with vets, I identified a cluster of symptoms that existed in those with this desire to redeploy: poor adjustment to civilian life, anger management issues, unhappiness, difficulty in relationships, and lack of purpose. Their need to deploy seemed to reflect desires for a tribe, fame, machismo, a natural high, and a greater purpose.

Determined to figure out what was going on, I searched the internet for a possible diagnosis. No objective, empirical basis for a longing to return to combat existed that covered both those who had been in combat and those who did not, both of whom could suffer similarly. I needed to figure it out.

Eventually, I called this cluster of symptoms Warrior Withdrawal Syndrome (WWS): Baseline Adjustment due to Military Functioning (BAMF). Because at this point WWS is an unknown syndrome, these symptoms are sometimes diagnosed as chronic adjustment disorder, PTSD, combat addiction (a form of PTSD), or other trauma- or stressor-related disorders.

Increasingly, I felt the need to put together this information into a book. This need became more compelling after meeting with two buddies: one who lost both legs, and one whose body was intact but whose mind was equally, if not more, crippled.

The one who lost his legs was now a big deal. He got a medal and went on podcasts. He felt he had contributed. He garnered sympathy and compassion. The non-injured one, who had pulled the one who lost his legs out of a hole, should have felt like a hero. He didn't. He had no visible evidence of having served heroically and felt it was all for naught. He was obsessed with returning to combat to prove himself a hero.

His wife turned to me, asking, "How do I make him feel what he did over there was not nothing? There's no book to guide me."

Hmmm . . . the universe was talking to me. My goal became to write a book that would provide an understanding of WWS. Having a book for these vets to refer to is especially crucial because many initially refuse mental health interventions, feeling it would make them seem "crazy."

I hope this book will change their minds. Reading it might jiggle some change in mindset from "therapy means I'm crazy or broken and need to be fixed" to "therapy is the kind of training that I need to give me a reason to get out of bed in the morning."

In Part I, we will explore my personal journey with WWS and how it led me to devise a constellation of symptoms that, while might not present an official diagnosis, is different from anything out there describing the mindset of some vets post-military and that ties in with the mental health problems commonly seen in retired vets.

In Part II, we will explore how I negotiated war and family, how my wife and I initially "dated" through emails, and, once we got married and had a family, how my frequent deployments left me with little time to know, be with, and care for my family.

In Part III, we will explore how the military shaped my identity and others', how hanging up the uniform presented an identity crisis, including imposter syndrome, for many.

In Part IV, we will explore combat addiction, how it plays out as seeking machismo, order, tribe, a medal, purpose, and place in the world.

In Part V, we will explore therapy for vets to help them recognize and come to terms with WWS, providing tools to "retrain" so they can adjust in civilian life.

A disclaimer regarding women in the military. Throughout this book, I refer to men who served and not to women. Having spent most of my career in combat arms MOSs, I rarely worked alongside women. While I have had a few female veterans as patients at the VA, I do not feel confident that I can describe the nuances associated with being a woman in the warrior culture. The effect of WWS on women may be different than it is for men and is worth studying more in the future.

Let's begin our journey with my personal story of combat addiction.

PART ONE

COMBAT HUNGER

CHAPTER 1

PREPARING FOR COMBAT

I'm probably more comfortable inside a Marine Corps rifle company than I am anywhere in my life.

—Jim Webb

What makes someone wish to be in the military? For some, it's their only way out of a meaningless life. Many of the guys in my platoon had been orphans or were from broken families. They lacked skills and barely squeaked through high school. The future looked dim. The military was their ticket to a better life. They didn't have to become a mass shooter and release their frustration and rage with an AK rifle; they could kill legally and be rewarded as heroes. Others joined looking for adventure.

While the "dregs of society" dominate the military, recruits also come from the elite; think of all the West Point graduates. Guys like me, who came from stable families, had a college degree and a bright future, joined to serve their country. What united many of us was a reckless desire to live on the edge between life and death.

The military was in my blood. My dad, whom I greatly admire, grew up in Rio de Janeiro, Brazil. In the early 1970s, he came to the U.S. to practice medicine. At that time, Brazil had been under military dictatorship since 1964. My father saw his future in the land of the free.

A pathway to citizenship could be expedited if he joined the Army, and he did in 1976. He was accepted as a captain in the Medical Service Corps in the reserves. Given his years of active practice as a physician in the civilian community, he was immediately promoted to major. Stationed at Fort Detrick in Maryland with the U.S. Army Medical Research Institute of Infectious Diseases (USAMRIID), he continued research into infectious diseases. The fall of Saigon was in 1975, so he just missed having to serve in Vietnam.

My maternal grandfather had been a military officer in Lithuania. His military career was brief, as the Nazis ran through Lithuania quickly in 1941. Impressed that he had been a soldier, I would search in his house for his uniform and any medals.

From a young age, I was enamored with war. My brothers and I loved playing soldiers, me more than them. Our favorite toys were GI Joes, still popular to this day. We would run around the neighborhood in camouflage, rifles pointed at imaginary enemies or at each other, shouting, "Bang! Bang!"

In college, I joined the ROTC with the hope, initially, of becoming a Navy SEAL. That ended up a pipe dream because I have poor vision. Still, I was hoping to be part of something daring and to taste real combat. This finally came about in 2003 with the invasion of Iraq, but getting to that point took considerable preparation.

My first deployment was the 31st Marine Expeditionary Unit (MEU) in early 2001. A MEU is a quick reaction force, ready for immediate response to any crisis, whether it be a natural disaster or a combat mission, often mobilizing within hours. I spent six months in the MEU, stationed around Okinawa as part of this response force.

My second deployment was a four-month rotation to Kuwait that same year, with a crisis management command as a ready force following the terrorist attacks on September 11, 2001. We were a command staff set up to manage nuclear, chemical, biological, or high-yield explosive attacks in the area. Concerns about Iraqis potentially destroying the Kuwaiti desalinization plants or flooding the Persian Gulf with

Iraqi oil would require a quick and coordinated multinational response. We were trained to be that command. None of those attacks or accidents occurred.

All in all, these first two deployments disappointed me. Neither involved hostile action, and my desire for combat was unsatisfied. Further, my company commander from 2000 to 2001 was god-awful. Donald Marone was a self-serving, short, "skinny-fat" guy with an apparent Napoleonic complex who tried to prove his superiority by treating his officers with little regard. Most of the time he was sitting in his office instead of being in the field with the troops.

When we trained, I, as a platoon commander, would have to be the range safety officer (RSO), and the executive officer, Jason Reed, would have to be the officer in charge (OIC). The RSO does not participate in the training and is responsible for safety only, like a referee. The OIC is responsible for the training and can participate, like a player-coach. Both roles bear responsibility for making sure everything goes well. In this way, anything that might go wrong—an injury, a lost weapon, and so on—would be my and Jason's responsibility, not Marone's. Among other things, Marone forgot to write the reports on his lieutenants, including me. Missing reports meant that we might not be promoted.

His ineptness knew no end. In one training session, we prepared for a training mission in Guam, where we would be inserted by helicopters. Man, those birds roared! Each helicopter could carry twelve Marines, and consequently, we were sorted into groups of twelve. Because only four helicopters could load at once, only forty-eight Marines would be inserted on the first wave, and only twenty-four at a time for each wave because only two helicopters could land at once.

Our numbered groups of twelve were called "helicopter serials." I was a platoon commander in charge of about twenty-eight Marines, with some extra Marines for the mission. The assault element is the platoon tasked with destroying the objective. Knowing the serial numbers, I suggested to Marone that his serial be bumped from the first wave so that I could bring my full complement of troops in those four helicopters.

"I think we should change the bump plan," I said. (The "bump plan" referred to which serials would be left behind in case an aircraft malfunctioned.)

"Which serial would you change?" he asked.

"38202," I responded, referring to Marone's serial.

Marone checked the manifests and said, "That's my serial! You can't bump me!"

"Sir, I need the assault element to land all at once so that we can move to the objective. Your serial means I have to wait for the second landing before moving toward the enemy."

"But I need to be there."

"Sir, I have all of the assets that I need to take down the objective if I have all three squads. Your serial doesn't provide me any combat power."

"But I have the radios to the battalion!"

"Good point, sir. Will you attach your radio operator to me so that I have the communications to the battalion?"

"Oh, de Sa, you're so funny." "de Sa" was my nickname.

I tried not to laugh. I was so sick of his bullshit.

For the entire deployment, Marone was a glorified clerk who would get lost leading the company in the mountains of Camp Pendleton, which happened several times during training missions. His commander had witnessed Marone alone in the office while our whole company was in the jungles of Okinawa training for two weeks.

Discouraged and sick of taking command from a moron, I wanted to switch to another section of the Marines. I had longed to be in Force Reconnaissance, at the time the closest thing to special operations in the Marines. Their motto being "Swift, Silent, Deadly," Recon Marines do much of the same training as Navy SEALs, which I had originally hoped to be when I first joined the military. The training to be a Recon is grueling, and about half of the recruits don't make it. I felt I was physically fit enough to make it.

In 2002, before Iraq, I'd been offered the opportunity to go to 1st Marine Division Reconnaissance Battalion as the headquarters and service company commander. But there was a catch. Though I would be elevated to the rank of captain from first lieutenant, I would not be in a deploying role. In 2002, and generally since Operation Desert Storm in 1991, 1st Recon did not deploy as a large unit. The platoons were attached to the division infantry battalions ("infantry" being the ground troops that engage with the enemy in close-range combat). That meant that within 1st Recon, an elite unit with less than four hundred Marines, only the platoon commanders would deploy. If I accepted the assignment, I would instead be a support leader and not a platoon commander. Since I was chomping at the bit to deploy, I chose to go back to a regular infantry battalion that was deploying.

In 2002, I came back from Kuwait to Camp Pendleton in California and became a company executive officer in preparation for deployment. Having served as a general's aide, I had some choice as to where I would go within the division. I chose to join the 2d Battalion, 1st Marines (2/1), who were scheduled to go on an MEU.

They needed an executive officer for the small boat infantry company. The boat companies train in Coronado, California, and become proficient in using combat rubber raiding craft (CRRCs), which carry up to eight men, as their method of insertion from the large Navy ships to the shore.

The CRRCs are more clandestine than the lightly armored, fully tracked assault amphibious landing vehicles (AAVs) and the large helicopters. Since carrying the boats and moving them from shore to objective and back again required strength and above-average swimming, boat companies typically had a more physically fit Marine. I fit that bill. Not only was I a good swimmer, but I was also strong. Furthermore, I had been a platoon commander with the boat company in 1st Battalion, 5th Marines (1/5). This would be an ideal placement.

I knew some of the other officers at 2/1 from my time in officer candidate school (OCS) in Quantico, Virginia, in 1998, and the basic school (TBS) in 1999. One of them, my friend Kevin Stern, warned me that my commander, Rodney Clay, was the commander from hell.

"Get ready," he said. "The fuckin' asshole knows squat. He's dangerous. Don't mess with him." To avoid being under Clay's command, Kevin had moved out of the boat company.

I thought Clay couldn't be as bad as Donald Marone—*ha!*

Around June or July 2002, I returned to Coronado for more training. Previously with 1/5, I had completed the scout swimmer course. This time, I would complete the over-the-horizon navigation course.

My former executive officer and mentor from 1/5, Jason Reed, was in charge of these courses. An aggressive but sensible guy with a wide, muscular trunk and piercing green eyes, Jason was a Recon Marine. He, too, warned me that Clay was an asshole.

After a month of training, Clay arrived with the remainder of the company as the new commander to complete the small craft raid training at Coronado.

A small guy with a meek voice, he asked my name.

"Varpas de Sa Pereira, sir."

"What kind of fucking name is that?"

"Portuguese, sir."

"Yeah, well, you're in America. You should fucking change your name. That's too hard to say. I'm going to call you something else."

"Yes, sir."

Disappointed with some of the Marine's efforts, Clay made us trainees work past the requirements of the course. Jason would have to pull us from events because we stayed past the scheduled training time.

"Goddamn asshole," Jason said.

"Jesus, dude," I said.

When the trainers left, they took their rigid hull inflatable boat (RHIB) used as a safety vessel with them. Presumably, without a safety vessel, we wouldn't be able to keep training.

"Fuck that," Clay said to me. "You can be the safety boat. These jackasses need to keep training. We are going to keep doing this until they get it right." Either we were too slow in disembarking the boats, or our formations looked sloppy, or some other part of our movement wasn't good enough, and we kept doing it over and over until he was satisfied.

For two weeks, we had a lot of late nights. Jason would say good night while we were still training on the beaches and the water.

Once combat started, I was the second under Clay's command. What a cluster fuck that was! An inept coward with a tiny ego, Clay had to be continually stroked. Defy him even slightly and you were on his shit list. I was at the top of that list, and it cost me big.

CHAPTER 2

ON THE WAY TO WAR

*In my experience, Marines are gung ho no matter what. They will
all fight to the death . . . they're badass, hard-charging mothers.*

—Chris Kyle

As early as April 2002, there was talk of invading Iraq. Then the invasion didn't happen, leaving us Marines antsy and frustrated. Finally, in the last couple of months of 2002, we were issued brown desert boots to break in to get ready to deploy. Having failed to experience combat in my first two deployments, the thought of going to war stirred my blood.

In the meantime, Clay's incompetence echoed loud and clear, even before we were officially at war. On one of the last few large-scale exercises, we were supposed to be in our gas masks for a lengthy time, simulating a chemical attack. Clay told us to take off the equipment because he didn't like being in his gas mask. We thought the exercise was over. Then other officers heard him calling his command and lying that we were still wearing the equipment.

On another large-scale training exercise, Clay broke the rules of the scenario, maneuvering the company behind the enemy to surprise them and make for a fantastic attack.

VARPAS DE SA PEREIRA, PsyD

He was supposed to come up with a scheme of maneuver to attack an enemy objective on the side of the hill. The enemy force knew we were coming in from the north and had oriented their systems in that direction. Attacking from the south would be a tactical surprise and demonstrate a significant victory.

The only way to get around the enemy to attack from the south would be to scale a large mountain ridge or cross Interstate 5 as it passes through Camp Pendleton, march two hundred Marines with machine guns and mortars through the San Onofre State Beach parking area, and then come back onto the base south of the objective.

While this resulted in a tactical surprise, we broke the "*Top Gun* hard deck." For this scenario, the I-5 served as the "Pacific Ocean," and we weren't supposed to cross it. Further, we may have illegally transported those weapon systems off the base and then back onto the base.

Fortunately, nothing happened as we marched the miles down the state beach parking lot while surfers and campers stared at the endless column of Marines with weapons passing by.

Things were so bad with Clay that the officers and Marines of the company developed an acronym FUBOC, for "Fucked Up by Officer Clay."

"Fuck!" I said to myself. "And I have to be under Clay's command during combat." I had no idea how bad this would prove to be.

In January 2003, after four years of training in infantry and two non-combat deployments, my company—part of the infantry battalion landing team (BLT) for the 15th MEU—was assigned to be a ground combat element for the 3 Commando Brigade of the U.K. during the initial invasion into Iraq. We set sail from San Diego for Kuwait under the command of Colonel Worthington. Finally, I would get my chance to look the enemy in the face.

Our intent was to invade Iraq and defeat Iraqi military and paramilitary forces. I was a first lieutenant, the executive officer of an infantry company and second in command to Clay, heading a group of about 180 Marines.

To prepare us for combat, our infantry battalion landing team would be reinforced with an artillery battery, amphibious assault vehicle platoon, combat engineer platoon, light armored reconnaissance company, reconnaissance platoon, and other units as the mission and circumstances required after we landed.

I was charged!

Thoughts of war with Iraq began with President George W. Bush arguing in 2002 that the vulnerability of the United States following the September 11 attacks of 2001—combined with Iraq's alleged possession and manufacture of weapons of mass destruction (later proved erroneous) and its support for terrorist groups that included al-Qaeda, the perpetrators of the September 11 attacks—made disarming Iraq a priority. UN Security Council Resolution 1441, passed on November 8, 2002, demanded that Iraq readmit inspectors and comply with previous resolutions. Iraq appeared to comply with the resolution, but in early 2003, President Bush and British Prime Minister Tony Blair declared that Iraq continued to hinder UN inspections and that it still retained proscribed weapons.

From the moment we set sail for Kuwait, we had a hunch an Iraqi invasion was inevitable, as the brass told us to get to the Middle East *fast.* We made it to Kuwait in late January after three weeks at sea. Going from San Diego to Hawaii took seven days, followed by another week or more to get to the Far East. We resupplied only once, in Singapore, where we stayed for thirty-six hours. Other refueling and resupplies were done "underway"—while we sailed, a supply ship came alongside and then ferried over supplies by aircraft, with cables between the ships, or with small boats.

On the ship to Kuwait, we had little to do. Only limited training could be accomplished in a tiny amount of space since the flight deck was taken up by constant helicopter and AV-8B Harrier jet training. Instructors would march junior Marines on the flight deck in the Corporal's

Course, a distance education program to provide students with the basic knowledge and skills necessary to become successful small-unit leaders, using realistic problem-based situations that a Marine corporal will encounter.

To keep our shooting skills sharp, we would fire our weapons off the flight deck at targets. Sometimes the targets were the biodegradable trash bags tossed over the side of the ship. Other times we would set up paper targets on stands at the edge of the flight deck. These live-fire training evolutions had to be coordinated so that none of our projectiles would hit another ship in the MEU.

To fill in time, I worked out in the tiny gym, ran on the flight deck, and watched movies on a computer after I had read the books I had taken and, of course, all the *Penthouse* magazines on deck. I would check on the Marines below deck and play video games with them or my peers. I had enough time during that sail to complete the training needed to wear a green Marine Corps Martial Arts Program belt. Mostly, though, time dragged.

Once we arrived in Kuwait, however, a different kind of reality ensued: boredom interspersed with danger.

Stan Korhonen—the company gunny sergeant who is the senior enlisted tactical advisor to the company commander—and I had flown into Kuwait in advance of the rest of the company to scout out our section of the camp. Getting off the plane was like stepping into an oven. My hand flew over my mouth as the hot wind blew tiny granules of sand everywhere, even into my throat, eyes, ears, nostrils, and hair.

My battle buddy and driver Stan was a no-nonsense, short, gangly, dark-haired guy in his thirties from the middle of Nowhere, U.S.A. A former sniper, he had a dark sense of humor that kept me entertained, like duct-taping an Ozzy Osbourne bobblehead on the hood of our vehicle.

We set up "Camp Bullrush" on a square of desert north of Kuwait City, with a berm, or earthen wall, around it—our living space for what we thought would only be a few days. As it turned out, for most of

February and March, our company hunkered down in our small, two-man, putty-colored tents in the middle of the vast, barren desert, the boring vista dotted only with small earthwork hills. Warm winds rattled the canvas as we sweated and wiped the sand from our bodies nonstop. We stunk like hell.

When the rest of the Marines arrived, they brought with them one of the two Humvees the company had, a large High Mobility Multipurpose Wheeled Vehicle (HMMWV) with doors and roof made of canvas. They also brought one of the small boats, a CRRC with an engine to maintain, just in case.

"In case of what?" I asked the assistant operations officer.

"I don't know. Maybe for using them on the rivers of Iraq."

"Whose fucking idea was this? We can't wear armor on them. We'd just be sitting ducks to shoot at from either side of the rivers. Plus, there's a high likelihood that we won't have the fuel mixture for the engine and no real way to maintain it."

Fortunately, the folded-up CRRC and its engine were taken back to the Navy shipping. We would need the space in the back of the Humvee for our ammo and supplies.

Small metal towers were placed in the corners of our compound to serve as the lookouts for anything emerging out of the desert, manned by stern-faced Marines with large-caliber machine guns and Humvees. Being a lookout was a boring job, and we Marines dreaded it. During the day, we stared out at 360 degrees of sun-drenched horizon around the command post, sweating bullets. During the night, we stared out at total blackness and froze our asses off, shivering as the temperatures could drop to twenty-five degrees Fahrenheit.

To many, placing these towers seemed silly. Still, we had to take precautions; the year before, in October 2002, a couple of terrorists had attacked the boat company training on Failaka Island in the Persian Gulf. Because those Marines were there for training, they had no live ammunition. Out of 150 members of the company, the only people who had live ammunition and could defend against the attackers were the sentry who guarded the blank ammunition and some of the staff's NCOs, who had pistols.

The gunmen injured two Marines, with one of them dying in surgery from his wounds. The Marines killed the two gunmen, and then, with a resupply of ammunition, tried to find a third suspected gunman. They detained over thirty-one locals and turned them over to the Kuwaiti government for questioning.

This event led to some policy changes for the Marine Corps. Any event with a large group of Marines, from training to celebrating the Marine Corps birthday at a gala, would be required to have "guardian angels," armed Marines to provide security for everyone else.

As we waited for our marching orders, we spent our day mostly digging fighting holes, filling sandbags, or taking orders from an idiot like Clay because he outranks you.

Often, we were hit by sandstorms, know there as "shamals," where winds would gust up to fifty miles an hour, sometimes shredding the canvas tents and covering us in sand. Many of us walked around with runny noses and inflamed eyes from the constant dust.

Life in the desert consisted of the bare minimum, a bit like going camping. Most of the Marines had to share a tent, though, as an officer, I had my own. I would have been fine sharing one. As one of four brothers, I had always shared sleeping arrangements growing up. The close quarters of military life seemed natural to me.

We chomped dinner out of plastic MRE bags in the evening swelter, faces and bodies sunburnt from the sun baking down all day. A pack might include four different entrées: beef stew, chicken fajita, spaghetti with meat sauce, and vegetarian chili. With each entrée was a side dish, a snack, and a dessert, as well as a drink mix and utensils.

The meal required no preparation or refrigeration—not exactly mom's cooking, and a lot of us lost weight, pulling up our pants without buttoning them.

Stuffed into our MRE bags was an array of junk food, like Skittles, Pop Tarts, and Slim Jims. We would tear through the bag for the goodies, a process called "ratfucking." This was also the term used when someone opened a new case and picked through all the meals for their preferred one. You were supposed to be happy with whatever you pulled out—in other words, you get what you get.

But some discarded the meal and just ate the junk, which they gathered into a ratfuck box, keeping it in their Humvees. I ate both the meal and the snacks. However, within a few months, I wound up eating only one meal per day (you're given three), otherwise surviving on the crackers, cheese, and peanut butter in the ratfuck box.

Sometimes, we got care packages with home-baked cookies that we devoured. These packages took weeks to arrive. If not packed well, the cookies would be brittle and crumbled.

Eventually a larger white tent was set up for cooked meals, with folding chairs and tables next to the tent. We had no clue where our food came from. We assumed locals had prepared it somewhere off-site. We would push a tray or paper plate and point to what we wanted.

Marines didn't want to eat these meals. The religious beliefs of the locals prohibited eating pork, and as a consequence, we had no idea what the sausages were made from. Long and skinny, we joked that we were eating "camel cocks." Though Marines greatly preferred their MREs over these sausages, we were ordered to eat in this makeshift cafeteria at least once a day. The higher-ups didn't want to alienate the locals.

Even worse than putting up with the lousy food was putting up with the lack of plumbing. We had no toilets or showers. Several months went by before I got a shower, and even then, it was from a truck designed to decontaminate vehicles.

Lack of sanitation and hygiene resulted in "Saddam's revenge," an outbreak of infectious diarrhea. At times, I squatted over a ditch dug to be our toilet and felt a nice breeze around my bare butt. If I looked down, though, I'd see that it wasn't a breeze—just a bunch of flies eagerly expecting what was coming next. There's little doubt that these conditions were keeping us ill.

However, this version of Saddam's revenge was minor compared to the Saddam's revenge we would experience later in An-Nasiriyah, where we fought a battle from March 23 to April 2. Likely the result of drinking contaminated water, vomiting and diarrhea rendered a third of my company combat-ineffective during the conflict. Our Navy medic corpsmen had to ration IV bags. Only those with the worst symptoms would get some relief from being rehydrated. The rest of us tried to manage the embarrassment, ranging from defecating in our pants to throwing up while sitting in the "bathroom area," a deep trench with metal pipes laid across the trench and a sandbag acting as a butt stop. This is to say nothing of the stink!

Sweating profusely, we needed to drink volumes of water to avoid getting dehydrated. Since Iraqi water was not well sanitized, we relied on either deliveries of bottled water or the purification of fresh water from the Euphrates. Still, there was never enough to replenish all the water we lost from perspiration. Many Marines opted to drink sugary soda in place of water, but that sometimes led to dehydration. They would buy these small twelve-ounce bottles of soda from the local kids for a U.S. dollar. Local Iraqis weren't interested in the local currency, the dinar that still bore Saddam Hussein's face.

The kids would open the bottle immediately with a bottle opener and expect you to drink the whole bottle right away. These glass bottles were a precious commodity for them, and they wanted them back. They would fill the bottles back up and re-cap them to serve again.

Even though Kuwait seemed safe, complacency was a luxury we couldn't afford. Surrounding us were people who stared at us with daggers in their eyes, ready to kill us. Under constant threat of car bombs, roadside bombs, suicide bombers, and mortars, we could never be off our guard. Is that vehicle near you going to blow up? Does that kid's backpack contain a suicide bomb? Is that trash that blew to the side of the road, or a deliberately placed improvised explosive device (IED)?

Local mistrust was understandable. Part of combat operations involves the intimidation of other combatants. When we got to Iraq, we reduced intimidation in secure places by taking off body armor and removing our sunglasses while talking to locals so they could see we didn't have the evil eye.

We didn't know when we might be called to action. In the military, an environment that is impossible to predict and always shifting is called "VUCA" (volatile, uncertain, complex, and ambiguous). That was our living reality. Would we invade Iraq? When? Where? We had no clue.

Hunkered down in the middle of the desert day after day, hot, sun-burned, soaked in sweat, constantly thirsty, often sick with diarrhea or dehydrated, and itching for action, we were riled. To blow off steam, we would fight each other as sport to release stress and keep in shape: arm wrestling, judo flips, strong-armed headlocks, lying in wait and then swinging out with knives and nicking our buddies, and "turtle fuck-ing," or slamming our helmets into each other. Some men played with weapons, putting a pistol to someone else's head for fun. At one point in An-Nasiriyah, some of the older Marines made two of the younger Marines box blindfolded in the middle of our compound. War became entertainment.

During the day, our aggressive energy took on a serious tone. While the higher headquarters were preparing for the war and designing the attack plans, we trained and rehearsed immediate action (IA) drills, including what to do when you make enemy contact or what to do if a vehicle breaks down.

I focused on running our IA drills and preparing the contingency plans for moving around in Humvees. This included everything from jumping in and out of the trucks to reacting to enemy fire to filling sandbags for hardening the trucks against direct small-arms fire.

We simulated a broken-down truck that the Marines would hop out of, moving to their assigned places in accordance with the "bump plan." If we ran out of imaginary seats for the Marines from the disabled truck, then the bump plan would indicate which of them would be left behind so the mission could progress without delay. The abandoned Marines would be picked up later.

Advancing on an imaginary enemy position behind a bulldozed sand berm was another small unit tactic we practiced. Berms are man-made piles of sand or dirt and dominate the Iraqi terrain. Built as walls, they are used as pasture enclosures, as well as windbreaks and military fortifications.

To be ready for unexpected contingencies, Marine observers shouted new targets and the names of simulated casualties. Like children playing soldiers, we shouted *Bang! Bang!*, mimicking fire from our weapons. We would advance on the berm, periodically taking cover face-first in the fine, powdery desert sand, getting down and making noises like we were shooting at the enemy—*budda budda jam!* While this wasn't the real thing, it gave us a buzz. We could release our aggression without danger.

We also rehearsed the scheme of maneuver. This was like football practice: you draw the play on a chalkboard and then walk through it in full gear so that on game day, the coach could call the play from the sidelines, and you'd execute it smoothly.

When not rehearsing and practicing maneuvers, we would horse around. I have photographs of "tough" Marines building pyramids in the sand like cheerleaders. Others searched for lizards and tried to keep them as pets.

A sailor from our MEU sent some Victoria's Secret women's under-wear to one of the Marines. The senior Marines decided that the junior Marine of the company was "required" to wear them, along with his minimum required protective gear. Naturally, the Marines took photos of this young man in a black thong with his hat, rifle, gas mask holder, and combat boots—and nothing else. Even the lieutenants got in on the banter, arguing whose ass looked best in the underwear.

"PFC Jones has the nicest ass in the platoon."

"Not just the platoon, probably the company!"

"I'd recognize that butt anytime."

"Nah, I think Lieutenant Corporal Peters would make those look better."

Later, when the Marines had to dig fighting positions—six feet by 2.5 feet, and chest deep—some added "carvings," mostly of penises. Other sculptures were like the sandcastle version of the *Kama Sutra*. These antics relieved the anxiety and boredom of waiting to go to war.

In March 2003, we left our quasi-protected headquarters in the middle of the night, lights off on the Humvees to not be seen, and moved forward stealthily to a new staging area, a tactical holding place about twenty kilometers south of the border. From there, it would be less than an hour to the border. Expecting to get little sleep here on in, we downed Nescafé instant coffee crystals and popped ephedra.

Holy shit! Combat would shortly become a reality. I held my breath in anticipation.

While sleeping in tents had not been living in the lap of luxury, now we had to hunker down and sleep in holes we dug in the sand with pickaxes, open to the air and near our defensive fighting positions. This placed us dangerously within range of the Iraqi weapon systems. Furthermore, this area was flatter and a more open space in the desert, lacking the berms around it for protection from direct-fire weapons.

A pair of helicopter gunships rattled over us, presumably flying north on their way to battle. Almost in unison, we pumped our fists in the air and shouted, "Yeah! Get some!" "Get some" is the universal Marine Corps cheer. The very heart of WWS, "get some" symbolizes the almost sexual thrill of war that every Marine hopes to experience.

Being "on the road" meant we were more at risk. Lacking phone lines, Wi-Fi, or radios, we were left unplugged and mostly in the dark. For access to the news, we relied on embedded reporters and what they were hearing, or the occasional brief from the higher command. This information was our only means of interpreting whatever policy decisions were being made.

Our embedded reporters and photographers had satellite phones. Occasionally, if a Marine had built enough rapport with the reporter, they'd let a Marine make a call home. Typically, this was frowned upon by commanders as there was no way to allow for equitable use of the equipment, nor could the commanders monitor what information was being relayed home. A Marine innocently telling their family operational details could compromise the security of the unit.

The reporters got their "marching orders" from their publications over these phones. Their lifeline to the world, the satellite phones enabled them to let us know what the major muscle movements were from the coalition governments—the U.K. and Australian governments at first, and then later most European countries. Because we had a Reuters and a BBC reporter, a fair amount of information came through a British lens.

Assuming it was only days before we would cross the border into Iraq, we were handed brand-new chemical suits. Since landing in Kuwait back in January, I had been wearing the same tan uniform. This was the first time I put on a clean uniform in two months. Although we did our best to keep these desert camouflage uniforms clean, inevitably they got dirty and, without access to laundry facilities, couldn't be cleaned. We had to keep our extra set in our packs in case of an emergency. We were expected to just be dirty and clean ourselves with baby wipes.

Our mission was to capture and clear the port of Umm Qasr, Iraq's only deepwater port, surrounded by a densely populated town. It was the only large port with direct access to the Persian Gulf, where most, if not all, of the naval shipping came into and out of Iraq. The modern port had a large dock and a channel that could handle oceangoing ships. The bulk of the Iraqi oil would be pumped to two oil platforms in the gulf.

In our company were four infantry platoons, including three line platoons and one weapons platoon. The "line" platoons comprised three squads of thirteen Marines and, because we were going to war, one Navy medic, or corpsman. Squads comprised three four-man fire teams led by an E-5 sergeant, the lowest sergeant in command. However, typically we didn't have enough E-5 sergeants and would rely on strong E-4 corporals to lead these squads.

The infantry platoons, with twenty to fifty Marines, were packed onto large trucks hardened with sandbags. Ahead of me were all four platoons in eight trucks, with Clay, my commander, leading in a Humvee with his fire support team (FiST). My position was at the tail of our convoy. As the second in command, I was like the second brain of a large dinosaur in charge of reporting to the front of the convoy where Clay monitored our progress.

This was our formation for driving to the north side of the port where we would dismount and clear north to south, with each platoon responsible for a different section of the port.

On March 20, more than twenty-five thousand Marines, twenty thousand British troops, and thirty thousand U.S. Army soldiers were loading their rifles and dropping belts of ammo onto feed trays, preparing for the "go" signal to move out.

As the motorized company, we wouldn't be the first to cross into Iraq. We would follow the "mech" company, designated Echo Company (we were Fox Company, according to the military phonetic alphabet for unit names), which included a tank platoon and lightly armored AAVs, as well as the light armored reconnaissance (LAR) company. However, we would cross in before Golf, the "helo," or helicopter, company.

We planned on crossing the border well west of the port and its neighboring town. Some farms west of the port were largely unpopulated, and we could likely evade some of the enemy that way. We would cross to the east of Safwan Hill, which commanded a good viewpoint of the border. Reconnaissance and sniper units would oversee clearing that hill to facilitate our crossing.

Once we crossed the border and the farms, Umm Qasr was wide open, brown, dirt desert again. Lacking terrain features, the city had a railroad bridge at the northern end of the port which provided a line of sight into the town to the west, which was hundreds of meters in the distance.

Moving into town was not part of our plan. The initial attack plans opted to push far and fast, avoiding the town centers along the way and getting to Baghdad as fast as possible. As our MEU was assigned to the U.K. command, we would work on the eastern flank of the assault, securing the ports along the large canal from the confluence of the Tigris and Euphrates near Basrah. While we were securing Umm Qasr, Echo would continue north, and the Royal Marine commandos would clear the Al-Faw peninsula.

After crossing the border, we would move north for a short way and then turn east to the port facility. The idea was to outflank defensive positions and then clear the port facility in the new port to the north, and then clear all the way to the old southern port. We would surprise the enemy by enveloping them from the west.

Although our intelligence section, S-2, still thought that the majority of the Iraqi soldiers had fled the area, we were expecting some fight.

This was going to be it, I thought. *No turning back. We were going to war.*

CHAPTER 3

ENTERING ENEMY TERRITORY

The die is cast.

—Julius Caesar upon entering the Rubicon

On March 17, seeking no further UN resolutions and deeming further diplomatic efforts by the Security Council futile, Bush declared an end to diplomacy and issued an ultimatum to Saddam, giving the Iraqi president forty-eight hours to leave Iraq. When Saddam refused to leave, U.S. and allied forces launched an attack on March 19, when U.S. forces finally engaged the Iraqi forces on Kuwait's border with preparatory fire.

The attack began when U.S. aircraft dropped several precision-guided bombs on a bunker complex where the Iraqi president was believed to be meeting with senior staff. Following that attack, a series of air strikes was directed against government and military installations.

In the first days of the war, U.S. and Australian special forces staged night raids on Iraqi command and communication posts, hoping to stop the use of weapons of mass destruction. U.S. intelligence reported signs that artillery shells with chemical or biological warheads had been distributed to Republican Guard units ringing Baghdad days before the war began, with orders to use them if they faced defeat.

In case rockets with nuclear, biological, or chemical components flew in our direction, we received chemical detection equipment—but no training on how to use it. I had one of those machines you see at the TSA security screening checkpoints at airports now, where they swab your bag and then put the swabs into devices to detect chemical or biological agents. Without training, we just turned on the machine, and it would "sniff" the air and give us results. Days later, the chemical officer for the battalion told us we were using it incorrectly. We'd been misreporting all-clear calls from the machine. There was a sigh of relief we hadn't all been incinerated.

Alerts regarding several inbound rockets would come over the radio, and we had to put our masks on until we got an all-clear message. Countless videos during training about the effects of nuclear, biological, or chemical (NBC) attacks filled us with dread. Skin blistering or melting, eyes burning, blood bubbling, convulsions, death—Hiroshima-like images streamed through our heads. *Shit!* Was this what it meant to serve our country?! Living day to day with thoughts of chemical warfare as our fate?

In some cases, the Scud missiles went over our heads toward Bubiyan Island in the Persian Gulf, where some of our artillery was located. None of us saw their impacts, so the donning and removing of gas masks seemed tedious.

Around March 20, we got our marching orders. Early the morning of March 21, we were in position to cross into Iraq. We trekked over a bridge that the engineers had built to cross a massive ditch. That was the border. Rehearsal was over!

My gunnery sergeant, Stan, drove our cargo-variant Humvee. His job was to manage the logistics, getting "beans, bullets, band-aids, and batteries" to the Marines who might need them. My job was to execute whatever the commander's operational plan was and to provide oversight, such as directing the fires of machine gunners and mortarmen, as necessary. The FiST's job was to use the radio for indirect fire assets.

Of course, once we crossed the border, many of these traditional roles shifted depending on what events dictated.

As an executive officer, I was in the vehicle commander spot (passenger side) of the two-seated vehicle, as Marine officers are not allowed to drive. Stan would be my driver for our whole time in Iraq.

Crammed into the Humvee were the novelty explosive rockets, tens of thousands of rounds of ammunition, C-4 explosives, tear gas grenades, and extra Javelin missiles. We also towed the demolition equipment of the engineer squad. At the last second, Stan and I added a sandbag or two beneath our feet were to help mitigate any landmines we might hit.

"One misplaced round," Stan said, "and all the explosives we are carrying will detonate. We'll be a pink mist."

"Fuck," I said. "If I lose my legs, shoot me."

"Roger that."

Added to the ammo were a spare tire, a tool kit, military food rations, several gallons of water, diesel fuel for the first few days of the war, and more. Stan and I were wearing helmets, vests, MOPP suits, and rubber boots. We also carried gear like a tourniquet in case a bullet hit an artery in a limb, as well as a roll of kerlix gauze to stuff into an arterial wound to apply pressure so we would not bleed out before the medic arrived. Bulked up like this, we could barely move.

Warriors have different ways of dealing with their nerves. Some guys get silly and boisterous; some withdraw into a shell. I contemplated. Would I ever see my family again? Would anyone remember me in ten years? Still, my anxiety did not cause my attention to waver. My focus was on the task at hand and keeping my unit safe.

Stan shaved a bit off the edge of terror with nonstop jabbering.

"This is fuckin' it," Stan said.

"Yep. It fuckin' is."

"Good thing we brought a case of Cokes with us," Stan said.

Knowing I liked to drink soda, he had managed to get a twenty-four pack, which was between us in the Humvee.

"Let's get some!" he screamed at top of his lungs.

I laughed. While I shared his sentiments, I'm not a screaming type of guy. Setting modesty aside, girlfriends have described me as the strong, silent type.

Weighted down, I was hot as hell. Along with my tan Kevlar helmet, a four-pound, relatively light military helmet that provides resistance against most small-arm projectiles up to a .44 magnum, and a green flak jacket with a SAPI ceramic insert, I had to contend with my gas mask, an auto-injector of atropine and 2-PAM chloride in case of NBC attack, and a pistol, though firing my pistol was not my job.

The terrain offered no distraction from physical discomfort. A dismal, endless vista of off-white tones of mostly dirt and never-ending sand, with occasional green brush that some starved cattle grazed on, it reminded me of the dull appearance of a construction back lot that has nothing growing on it, just piles of dirt and rocks.

One thing in our favor: though the view was mostly flat, one large sand hill that another battalion was tasked with securing had enough elevation to observe most of the border.

Occasionally, the monotonous terrain was broken up by a few camel caravans, or flocks of sheep and shepherds wearing traditional red scarves with long, flowing dishdashas, moving at a slow pace and largely ignoring us. At one point, we came across some Bedouin tents and square mud huts, with women in long, black robes outside watching kids who were covered from head to toe like their parents. Some kids waved and yelled with a grin, "Americans!" I found that heartwarming. *Lord*, I thought, *let me never accidentally kill an innocent child or woman in the battles to come.*

We wore a green chemical suit, a stark contrast to the deep beige tone of desert terrain. Rather than the uniforms camouflaging us and making it harder for the enemy to see us and take aim, we were like the targets we'd practice on—dark silhouettes against a light background.

"Why don't they just send out a flare lighting us all up!" said Elliot Tanner, the 3d Platoon commander and one of my lieutenants.

"Hey. There's nothing to fear," said Robert Sams, the 1st Platoon commander. "We're so fucking far advanced. We've got better equipment. We're better trained. We're tactically superior."

"Yeah," I said. "Don't sweat it." According to intel, the battalion and the majority of the Marine Corps expected limited resistance when crossing the border into Iraq. "Probably just a few Iraqi forces there to defend the port. Those motherfuckers will probably surrender as soon as they see us."

"Yep," said Sams, "their weapons systems are old and dilapidated. Their training is nonexistent or limited. Probably not a round will be fired at us. Our 'attack to clear' will seem more like a parade into the port."

"I heard the Iraqi soldiers never even fire their weapons," I said. "They're given three rounds per year to shoot their AK-47s. The battle is going to be like when my older brother held my youngest brother at arm's length, keeping him out of reach as he flailed his arms."

Sams, a big guy, picked up Tanner, a small guy, and held him out as Tanner flailed his arms. We guffawed.

How naive we were. As it turned out, things didn't quite proceed according to these expectations, and on that first day, the enemy had a say in my continued existence.

CHAPTER 4

DEATH STARES ME IN THE FACE

War is death. If we are to engage in war, then we should have to stare it straight in the face and call it by its rightful name.

—Aaron Huey

On March 21, on the way to Umm Qasr Port, our convoy of twenty vehicles stalled west of our target.

I was irate. "What the fuck?! Why aren't we moving?"

"No idea," Stan said. "We're stalled here like sitting ducks."

Being in an open area made us vulnerable to direct and indirect enemy fires. My eyes kept scanning for anything out of place. Maybe we'd get blown up, or injured, or die. "Shit! We need to get the fuck out of here."

Behind us were journalists who had caught up to where we were stalled.

"Can we move forward?" one of them asked.

"Are you fucking kidding?" I said. "That's *the* front line. There's no good guys in front of us."

They ignored me and quickly drove ahead of us, beyond the forward line of troops (FLOT) where we had no way of protecting them. Combat correspondents want to be where the action is, not back in the rear.

"Goddamn civilians!" I murmured under my breath.

Still stuck in place and watching the press charge ahead of us, I felt frustrated and helpless sitting in a vehicle still within sight of the Kuwaiti border.

"This fuckin' sucks!" I shouted to Stan. I jumped out of the truck to pee on the side. Not my usual hose explosion—in the desert, you barely pee.

Why were we not attempting to gain momentum and maintain the initiative? Stan drove us to the head of the column of vehicles so I could find out what the hell the holdup was and offer tactical advice.

As soon as I got to the front of our convoy, the incoming artillery started as Iraqi forces outside of the port fired at our crossing site. The sky lit up in pink, red, and orange tones from the bombardment, and the air filled with deafening whistling, rumbling, hissing, and blasting sounds.

Fuck! They weren't as weak as we thought. Now sitting ducks in the open vast desert approach to the port, we were paying the price.

The artillery battery firing at us was to the east of the port facility. It was an Iraqi D-30 battery, or six 120-millimeter howitzers. These cannons can fire ranges up to fifteen kilometers, or nine miles. With the organic weapons systems we had in the infantry company, we had nothing with enough range to fire back, even if we could see them.

Artillery, like bombs from airplanes, comes from nowhere. And just like that, you're dead. At least in a firefight, you could shoot back at the enemy. Drowning in an AAV, crashing in a helicopter, stuck in a Humvee in the middle of the desert—there's *nothing* you can do. And now, just outside of Umm Qasr, not even within sight of the port, we were taking fire helplessly.

The attack sounded like a jet flying overhead, except it wasn't going overhead—*it was coming at us.* Adrenaline pumped through my veins, a mixture of terror and thrill. I took several deep breaths to calm myself. Three more shells followed in quick succession. *Boom! Boom! Boom!* Jesus! My heart pounded, my stomach twisted, my eyes were wide with terror. I felt as if I was staring mortality in the eye.

Artillery travels thousands of kilometers from their cannons. We could not see who was firing at us. Even if we did, we had no means to fire back. We would have to rely on our own artillery or counterbattery fire (artillery units that use radar to determine where the shots were coming from) and then fire at those locations.

"Fuck! Shit!" resounded all around me.

Jumping down to the ground, I laid as flat and small as I could and hid behind my helmet, my only protection.

Another round landed ahead of us. "Holy shit!" While the artillery was only firing at us sporadically, they came terrifyingly close. If a single round had landed on one of our vehicles, they could have killed an entire platoon of Marines—*my Marines*. The guys I was responsible for.

We were "troops in the open," and the tactical mix used by the Iraqis should have been HE-VT, or high-explosive proximity fuse. This meant that those shells would have exploded in the air and rained down shrapnel on us. No way we would've survived that. Luckily, the HE quick shells fired at us detonated as they impacted the ground. If those Iraqis had better training, shrapnel would've rained down on us, and I'd be in pieces. Instead, "HE quick" meant the shrapnel mostly flew over our heads while we were lying flat on the ground.

Finally, the screeching stopped. I jumped up and flew to my vehicle, a big grin on my face and hooting with excitement.

"You nuts?!" said Stan. "Who smiles on the battlefield?"

"Me. I didn't die."

I was reminded of something Winston Churchill had said: "There is nothing so wonderful as to be shot at and missed." So true.

"Rest of the crew?" I asked.

"They're okay, as far as I know."

"Holy shit. That was close."

"Yep. Fucking cool!" he said with his face lit up like a Christmas tree. He pumped his fists into the air, shouting, "CAR! CAR! CAR!" CAR, or Combat Action Ribbon, is the medal awarded to Marines who have been in combat.

Slowly, my adrenaline rush dissipated, and I felt survivor euphoria. My breathing returned to normal and my heart stopped pounding.

Fatigue set in. With dust, sand, and sweat covering my face and body, I collapsed on a sandbag inside the vehicle, images of getting ambushed, being gassed, and shooting endless rounds of fire into the bad guys racing through my mind.

"Enemy fire?" shouted Clay in disbelief when I informed him of the ambush. "It's blue on blue! We must have been in their impact area by mistake. That fucking idiot Ron March told the wrong location of a target to our mortar platoon." "Blue on blue" meant friendly fire from our own mortars, and Marine Lieutenant March was the fire support leader.

Can you get more fuckin' inept than this, dumbass moron?! Based on what I'd heard on the radio, Ron's reports, and what the artillery forward observer had said, it was clear that Clay was wrong about the barrage being friendly fire. The Iraqis were trying to kill us.

Over the radio, I had tactfully requested that he give the command for us to continue to move forward.

"Terrapin 6, this is Terrapin 5. We're still backed up to the border crossing. We need to keep moving. Over." "Terrapin 6" was Clay's call sign—"Terrapin" referred to his company, and "6" referred to him, the commander. "Terrapin 5" was the executive officer.

"Negative, Terrapin 5. We are going to hold position. Over." "Over" meant Clay would accept a response.

"Terrapin 6, we're still in the open and they can bracket us. Over."

"Negative, Terrapin 5, that was Regulators. Out." "Regulators" was the call sign for our eighty-one-millimeter mortar platoon.

"Out" meant the discussion was done. I had no more to say. I thought, *Did the motherfucker care if we were blown to bits?*

Later I learned that Clay was so sure the firing was from our side that he had threatened Ron, the FiST leader.

"If you ever call in another fucked up fire mission, I'll kill you." For added effect, he did this while aiming his rifle at Ron from the hip.

Meanwhile, the company vehicles moved in a little closer to help extract the FiST and the other Marines from the immediate danger area. As the truck began moving, I noticed the FiST leader was being left behind. I grabbed him by the deuce gear and hauled him into the truck.

We thought that we had moved far enough away from the berm to be in a safe position. We were wrong, and more rounds impacted us, not too far away. Clay had moved far from his vehicle. To save his ass, Stan rolled in during a pause in the fire and pulled him into our vehicle. I let him take my seat since he was out of it, glancing down and shaking his head, not knowing where to go.

"Who put that motherfucker in command," I muttered under my breath.

I grabbed the rifle he had left behind since I carried only a pistol, and jumped on the hood of the Humvee as we pulled out of the area.

Once we reached another area of relative safety, Clay, now more lucid, brought the FiST together and tried to figure out what had gone wrong.

"Sir, don't you think we should move forward with the attack?" I asked.

"Negative. Resume your position in the rear of the company."

"Sir . . ."

"You heard me. Remember your role. You're not in charge. We're going to wait to link up with a section of tanks before we progress further."

This jackass is going to kill us all, I thought. Still, I didn't argue my point further, knowing it wouldn't get me anywhere.

We stayed stalled in the same location, within sight of the border several kilometers away from our objective of the port facility, so Clay could work with the fire support leader and "figure it out."

"This wasn't friendly fire," said Stan, shaking his head in disgust. "If they fire again, we're dead meat. We need to move."

"Tell me. We're a sitting target. The motherfucker's a moron."

Indirect fire, whether artillery or mortars, is a lot like a robot playing horseshoes. If they hit the target once, their information would be dialed in, and a second strike would hit us again. If they missed, they could make the appropriate mechanical changes and be right on target with the next volley. As it was, the Iraqis only missed us by about fifty meters, an easy correction to make before firing again. Ideally, we should have moved in the direction of our target. Anywhere would be better than where they just hit.

Indeed, more incoming fire struck near us again, hitting our vehicle and piercing my laminated tactical notebook. Holy shit! That was a close call. I had just stared death in the face. Afterward, I took a picture of the notebook. Ultimately, that photo got me the CAR.

Surely Clay would get us moving now. Nope. He wanted to wait for a section, or two, of our tanks to join us to continue to move toward the port facility. As far as we knew, the enemy had nothing that could damage our tanks, so the tanks could advance without fear, and we would follow behind. But this also meant more waiting in the same impact area.

We were shaking in our boots. Miraculously, only my notebook was hit, not my body—again, a big smile on my face. In the midst of combat, I had made it out alive. The adrenaline was pumping.

I took my place back at the tail end of the company and waited for the tank section that Clay had commanded, meanwhile monitoring on the radio their linkup with our company. By the time the tank section caught up with us, two hours had passed with the company immobile in a targeted area. We moved forward about a hundred meters at a time as the combined arms assault teams (CAAT) engaged the enemy and took prisoners.

"Sir," I said to Clay. "I suggest we use the Marines to deal with prisoners so that CAAT can continue to push forward again."

"Denied. Remain in the rear of the column."

"Sir . . ."

He ignored me. I tramped away shaking my head, muttering "asshole" over and over. This entire time the company was in an ambush area of an unknown enemy force. Luckily, the enemy did not engage the motorized column with their heavy machine gun and indirect fire assets, or I may not have lived to tell the story.

Once we reached the crossroads, we took another thirty enemy prisoners of war. We stalled again to use the entire company to take care of the prisoners. They were a sorry lot, mostly young guys in their twenties, and a few had no shoes and blistered feet. They looked dehydrated, so we handed them water bottles, which they took while staring into space in a daze. A few were crying. Some of the guys who smoked offered them Marlboros, which they grabbed with trembling hands. After they were fed and given clean clothes, we would inter-rogate them.

I drove forward to tell Clay that we could not sacrifice the initiative again. I recommended we leave behind a platoon, or just one truck with Marines, to round up the enemy prisoners and clear the enemy position, leaving the remainder of the company to continue in the attack.

With a nod from Clay but no verbal confirmation, I drove back to the end of the column to pass the information on to the Marines since the radios were no longer reliable. Then I drove back to the front of the column to see why we were held up again when we were supposed to be moving forward.

Clay, the narcissistic coward, had quickly sought cover, moving his vehicle into a gully where he could not observe what was going on with the forward elements of the column. I remained at the head of the assault in our Humvee to give him situational reports via radio when possible, or face-to-face as I moved between maneuvering units.

After the CAAT team passed through a checkpoint, Clay drove to a safety position outside of the port and used his FiST as a security ele-ment around his vehicle. This defied his own order to enter the port with his lead elements.

As Clay was no longer in a position to direct his platoons, I effec-tively took control and passed the info onto him as he sat limply in the Humvee, incoherent and pupils dilated. I attributed this to shell shock and carried out his original orders to use two platoons to clear the northern port and one to gain a foothold for Company G in the southern port. *Useless bastard*, I kept thinking to myself.

I drove with the Marines of 1st Platoon to their blocking position to ensure they were set in correctly. Their task was to set up a position that would prevent or block any enemy from interfering with the clearing happening in the northern port. We were headed south, down "uniform road"—a tactical name given to a road that facilitates communications about locations—when the column was ambushed from the residential area. The Marines from CAAT and 1st Platoon fired back, but I could not see any enemy targets to engage. I maneuvered out of the incoming indirect fire to a better position and had the column move forward.

When the trucks got past the ambush area, we drove back up north to check on the Marines escorting the enemy prisoners of war (EPWs) into a makeshift circular corral of barbed wire.

When moving through the kill zone again, we were struck by indirect fire only a few meters off. Shrapnel and debris flew into our vehicle, showering us. It cut holes into the attached packs and my MOPP suit, which protected against chemical, biological, and even nuclear threats. My stomach churned, and a surge of fear shot through me. I wasn't wounded, but I was hit! We had to continue through the ambush area to get to a position of safety. *What a clusterfuck!*

I moved to the front to intercept the Marines escorting and clearing the first ambush site. I advised Clay to leave only a squad to walk these EPWs to the collection point, which I decided was going to be in the vicinity of checkpoint 11H. Checkpoints are used to help control maneuver units and describe our locations in a way that only friendly forces would understand. Even if the enemy intercepted our transmissions, they wouldn't know what hill or crossroad we'd labeled "11H."

At this point, the other Marines boarded their trucks and headed south. In one of the initial engagements, CAAT had hit a civilian vehicle by checkpoint 11H, collateral damage when CAAT had fired on a truck that had been bearing down on them because they believed the people inside to be the enemy. By this time, I had driven past the injured civilians three times. Seeing that they would die without medical attention, I ordered the first truck coming with us to leave behind its corpsmen to provide aid for the civilians. I relayed the information to Clay and heard a call for a medevac over the radio. This meant one of our Marines was injured and needed transportation. I took a moment to guzzle down some water that was hot as piss.

Because he had put himself in a defensive position in a covered area, there was no way for Clay to track the progress of the company over the radio. What little he knew about what was going on likely came from the few times that the company gunny and I passed by his mini "fortification." If he had ventured forward with the troops on foot, as he had planned back in Kuwait, there would be some chance of injury. Safely tucked in the vehicle with the FiST surrounding him, he would survive.

I headed over to the entry control point that 2d Platoon had taken, picked up the injured Marine who was hopping on one leg, and drove him over to the casualty point. I then grabbed Marines who needed to be moved to the entry control point building to join their platoon.

I also snatched the human exploitation team (HET), or human intelligence Marines, to interrogate the colonel captured in the vicinity of the port's mosque. We moved into the new port with 2d Platoon to clear a landing zone (LZ) for Company G. The Marines then moved vehicles from the LZ in anticipation of the helicopter arrival.

From there I headed back to the casualty collection point (CCP) to inform the Marines and corpsmen manning the CCP that the area was clear and that they could move into the new port. I dropped food and water for the injured civilians. They needed transportation to be saved, so I commandeered an Iraqi vehicle at the port with keys in the ignition and had one of my corporals bring it over to the civilians, allowing them to drive themselves to a hospital. In addition, I got a light pickup truck to help us manage the transportation requirements over the large terrain.

I headed south through the ambush area to find out if 1st Platoon needed anything and to get a situation report. Once we arrived, I saw that 1st Platoon was in a good position. They requested to move into the old port to gain a foothold and to clear an LZ for Company G, the helicopter company. If they established an LZ there at the old port, then Company G wouldn't have to land north at the new port and march on foot through an ambush area. This way, they would be able to clear their target more rapidly.

I gave Robert Sams the go-ahead and moved back up north, encountering the vehicle with the injured civilians on the way to the hospital. One woman had been shot in the back, while another had a leg full of shrapnel wounds. They had a kid with them, staring into space. I couldn't tell if the boy was injured or not.

I escorted them through our positions and then headed back up north to relay our information to Clay.

At this point, I needed to establish if the primary entry control point (ECP) was secure. Driving forward, the area looked relatively safe, so I moved through and headed back south to gain Marines from 1st Platoon to assist in EPW security at the main ECP. They were still in their position, and the LZ was clear. Our company first sergeant called in the helicopters to evacuate our wounded, marking the LZ for them and keeping tabs on Marines we'd send back to the hospitals.

Since Clay had remained sitting in his vehicle in a secure position outside of the port with the FiST still providing security, I requested that Clay's vehicle be used as the on-call ambulance by the company first sergeant, according to the original operation order to vacate the vehicle and travel behind his advancing platoons.

"Sir, it's necessary to give up your vehicle in case someone is injured. The company vehicle won't be able to medevac another person, especially if the injury is critical."

"I agree," said Clay. "When a medevac call comes over the radio, I'll give up the vehicle."

"Sir, you should collocate with the company first sergeant, who will be receiving the calls for medevacs."

"Why can't the first sergeant collocate with me?"

"The company first sergeant is moving forward into the port facility to call in the helicopters."

"What the fuck. He's already moving in?"

"We cleared the area already."

"No shit. Okay. Time for me to move forward." He collected his security team and did just that.

I headed back into the port to find out if the helicopters for Company G were already called into the northern LZ. They were, and we could not divert them to the new LZ by Sams and 1st Platoon. I watched as some of the CAAT Marines engaged enemy forces in the vicinity of the police station. Then I headed back to the LZ in the new port to link up with Company G.

Clay was not around, so I gave the sequence of events to Bill Bradley, the commander of Company G. He asked about a more direct route to their objective, which I had discovered.

"Sir, I can lead them down that route."

"Roger that," said Bradley.

I led Company G through the main ECP. There I discovered that the buildings in the immediate vicinity were not cleared. I grabbed a couple of Marines and made a hasty clearing of them. Then Company G was allowed to pass, and I headed back to the company command post.

Once there, I linked up with the company gunnery sergeant and grabbed a five-ton truck to bring some EPWs back for security and to drop off supplies. As we headed down uniform road, we were flagged because there was a casualty with Company G. We grabbed the injured Marine, whose leg was bleeding profusely, and brought him back with us.

We turned around to head back to the blocking position (BP) 1, where 1st Platoon was set up defensively. On the way, we were flagged down for another casualty, an older Iraqi male with a bloody bandage wrapped around his head. This time we sent the five-ton back with the casualty and then finally headed down to BP1 to ascertain the situation.

Clay was there, well behind the blocking position, trying to conduct a passage of lines, an operation in which one unit passes through the lines of a stationary unit. He had called over the 1st Platoon commander. Once we arrived, I needed Marines from 1st Platoon to reinforce the main ECP.

By the time I found the platoon commander, they were not yet ready to conduct a passage of lines, so we headed back up north without the extra security. Clay, the fucking dumbass, ordered me to do nothing outside of watching the EPWs.

Like hell I would do nothing and let things fall apart! Clay was focused on the passage of lines and no longer leading the Marines of the company. It was up to me.

His approach to the passage of lines was laughable. Similar to the way he had unfolded his map in the middle of a kill zone to determine "what went wrong" with the assault across the border, he was now engaged in another map chat with the other commander.

After that first series of artillery fire we took at the border, Clay had called for all the officers to meet him at his vehicle. He unfolded his map and drew the firing lines of Regulators and the British Royal Artillery battery to show how Ronnie March had made a mistake and landed the fire on us.

In Clay's mind, that initial volley was all that we would have to take because we had caused it and now weren't calling for fire anymore. It didn't matter that Ronnie, Ron Jameson, and I had already said we hadn't called for any indirect fires. Our assertion was confirmed by the battalion operations center. This obtuse idiot tried to take a "time out" in the middle of combat to discuss "our failures" earlier!

Now Clay did a "map reconnaissance," looking at a map and talking through operations with the commander of Company G. He discussed with Bradley where the friendly forces were located and how to ensure that the other company could pass through our defensive positions without being fired upon.

The only problem was that Clay had no idea where anything was since he got his information secondhand through me without a "map reconnaissance." Either he guessed or simply told Bradley our Marines' location based on where they were "supposed" to be in accordance with his orders. If he now labeled BP1 as an enemy position, Bradley's Marines might open fire on our guys! This would be contrary to the point of a passage of lines.

The passage of lines should have been simple: 1st Platoon was oriented to stop the enemy from moving from the south to the north; Company G was moving from the north to the south. All it would take was for Company G to indicate when their last unit had moved south of 1st Platoon, and then 1st Platoon would reorient to prepare for their next mission.

This approach somehow eluded Clay. He couldn't explain anything to the other commander. And yet, because this event had high visibility, as his superiors would be in the area, he felt like he needed to be there "directing the traffic."

While he was engaged in this task, I remained in the vicinity of the EPW collection point where Clay ordered me to be, because with Company G moving south, that fight was over for us. Establishing our defensive position in the north would be the next major task.

We now had significantly more EPWs than Marines guarding them, and we were looking to establish a mortar firing point. Up to that point, our mortars had fired in a hasty fashion when needed. They were less accurate than if they could set up their aiming stakes, plot their position, and unpack rounds to be ready.

Because we were getting prepared for a defense, it made sense to find them a point where they could sink the baseplates that stabilized the mortar tubes and be ready through the night with explosive or illumination rounds.

While finding a good position, I heard calls for aid from the main ECP. I headed there to help out.

Our 2d Platoon needed more Marines to man the area. I found the 2d and 3d Platoon commanders to grab the extra Marines. When I spoke with the 2d Platoon commander, Bruce Dean, I realized clearing the entire port before dark was not possible.

"What's taking so long?" I asked him.

"Some of these things are filled with pallets and heavy machinery. They're not open and clear like we thought."

"So, at this rate, we're not clearing them fast enough to get you in a defense by nightfall."

"No, and some of these are taking longer to get into because they reinforced the doors with rebar."

"Have you taken any fire from any of these warehouses?"

"No."

Dean suggested that 2d and 3d Platoons, about sixty Marines total, do a cursory clearance, and then they could make it through the whole port.

"Makes sense to me," I said. "I'll let Ernie Thompson know." Ernie was the platoon commander.

I headed to the 3d Platoon's position to pass on this information and then headed back to the command post to get other Marines over to the main ECP. There were none available, so we headed back to the main ECP, a makeshift holding pen we had created in the port facility, which was the main gate to the port from the west. We grabbed the EPWs there and marched them back to the command post (CP) and EPW collection point.

Finally, once the platoons had cleared through the whole port, the gunny and I came up with a plan to grab all of the Marines with trucks and bring them to our position for a defense. The 2d Platoon was tasked with guarding the main ECP. The 1st and 3d Platoons would rally around the CP before punching out to their security positions. They would create a perimeter around the CP, which was at the northwesternmost warehouse where the EPWs were kept inside.

The Marines in the CP put up a radio antenna to increase our ability to communicate around the port. The platoons would take up positions, making a perimeter around the command post and the EPW holding area to fend off any counterattacks that might come through the night.

Once all of the positions were assumed and inspected, I came back and reported to Clay, who, in the middle of the defense, had been in the most secure location we had.

"Marines know their fields of fire and responsibilities for their section of the defense," I told him.

Essentially my comment conveyed that the Marines around him had a full 360 degrees of security and that the "fields of fire" approaches were covered. This was the most tactful way I could think of to tell him essentially, "I've got this," and to reduce his anxiety about getting killed.

I told him to go to sleep. He nodded off, and I managed the defense, relieved I didn't have to take any more dumbass orders from the moron.

Eventually, with supporting fire from the Royal Artillery and the British Army unit backing our movement with indirect fire, the Iraqi artillery position was neutralized and wouldn't fire at us again. There was occasional small arms fire from Iraqi soldiers, but no organized resistance. Later that day, one Marine was killed. Unfortunately, this Marine from Company E was killed by "friendly fire."

Once back at the base, I was too exhausted to eat. I didn't even roll my sleeping bag. I basically collapsed and tried to sleep. But I couldn't. I'd had my first taste of war, and it was intoxicating.

Without realizing it, I was now addicted to the thrill of combat. For the next seventeen years, I would be obsessed with redeploying to a combat zone, despite having the responsibility of a wife and eventually four kids. I was my own first full-blown case of Warrior Withdrawal Syndrome.

CHAPTER 5

MORE COMBAT

Some people spend an entire lifetime wondering if they made a difference in the world. But the Marines don't have that problem.

—Ronald Reagan

A s it turned out, our mission in Iraq was not quite over. On the morning of March 23, Iraqi snipers fired on a U.S. patrol near the town of Umm Qasr. While the patrol returned fire, Iraqi forces moved into one of the warehouses in the industrial area meant to store goods to be loaded onto ships at the port.

This didn't surprise me. I had predicted that upon entering Umm Qasr, we could run into enemy forces inside the warehouses, and I had suggested a strategy to my commanders.

The imagery of the port I'd seen at Camp Bullrush in Kuwait indicated the warehouses were large. I pointed to one of the ships docked at the port in the overhead image and said to the battalion operations officer, "That ship there is probably as large as the oil tankers that we see off the coast of San Diego. That means the new port facility is a couple of miles long, with warehouses that can fit a football field in them. Clearing warehouses that size and over that distance is probably the maximum that our company can do."

"Let me take a look," the operations officer said. He agreed with my conclusion, and the battalion staff shifted the plan. Securing the new northern port and establishing a landing zone for Company G would be our limit. Helos would fly into a parking lot at the northern edge of the port and then move through our lines to attack and clear the southern port. The force reconnaissance platoon would move through and clear the police station in the town.

To prep for a battle at the warehouses, I suggested to Clay that we practice our IA drills on a scale model in some of the buildings at Camp Doha, the main U.S. Army base in Kuwait, before we got to Iraq. "The buildings are about the same size as those in Umm Qasr."

"Battalion said no. It's too crowded."

"How about just the key leaders do a walk-through, so they can get a better estimate of what we'll see in Umm Qasr. Doesn't it make sense to let our small unit leaders, squad leaders, and perhaps even team leaders see an approximation of what their objective would look like?"

"Negative," said Clay.

"Aren't we supposed to train how we fight?"

We'd learned the importance of training. That meant everything from wearing the same gear we would wear in combat to practicing on terrain similar to that of the expected battle.

Fucking asshole glanced down at a paper on his desk and ignored me. I had to drop it. How frustrating. Not using every asset available made little sense, considering lives could be on the line within days. Instead, we just used fat, white ribbons on the ground to approximate the buildings, mostly relationally, as there was no way to practice on a one-to-one scale. We built our terrain models and walked through the plan over and over.

How dangerous could it be to clear the warehouses? That depended on whether they were empty, in which case there would be nowhere to hide, or filled with cargo or packaging that could provide cover and concealment—and be booby traps.

As it turned out, the U.S. patrol called in tanks that fired on the Iraqi positions inside one of the warehouses. It wasn't one of our patrols, and their calls for reinforcements didn't come over my radio. I saw some of the action as it was transpiring and had no idea who had discovered the enemy in that building.

These enemy forces occupying the building were about a kilometer from our perimeter. At that range, our M16s (an assault rifle used by the United States since the Vietnam War in 1963, based on the AR-15) were not useful. To avoid wasting ammunition, I stopped some soldiers from an Army unit attached to our MEU from firing at the building, as the maximum effective range of the M16 at an area target is about eight hundred meters and less for a point target. Our machine guns, mortars, and Javelin rockets could hit the target instead.

For a day or two, platoons exchanged fire with the forces inside. Intense machine gun and rocket-propelled grenade (RPG) fire directed at the U.S. Marines prevented us from moving forward to locate and destroy the targets, presumably inside the warehouse. We called in air support.

Two RAF Harriers launched airstrikes on the Iraqi-occupied buildings. They missed. Machine gun fire died down, but the Marines came under increased sniper fire from the town.

The fighting continued into the night, with the Marines unable to make any progress against the Iraqi forces. After several missions with specific targets carried out by SEALs and special warfare combat crewmen (SWCC), the building was leveled to rubble and the remaining resistance was finished off.

Later, I found out that some of these teams had established contact with the forces inside the building we were targeting. Not knowing friendly forces were in the vicinity of the buildings, we had fired mortars at the enemy in the hasty handheld mode. Of course, this led to another confrontation with Clay.

He had been in a meeting with the MEU command, and someone had asked about the mortars firing. Apparently, they were flying over the heads of those SEALs or landing close to them. Having no idea that we were engaging the building that way, Clay assured his commanders that we were not firing mortars. When he got back to the CP, he asked us about it.

"We're not firing any mortars, right?"

"We absolutely are firing mortars, sir," I said. "Lance Corporal Gomez deserves a medal for firing those things handheld and hitting the building."

"I just told the MEU that we're not firing any mortars."

"Well, sir, we were."

His head lowered. He was embarrassed. This would lead to significant repercussions, probably regarding Ronnie. Indeed.

"We hit the enemy building, sir—what's the concern?"

"They were landing in the vicinity of friendlies, you assholes!"

"What friendlies, sir? There's not supposed to be anyone west of the road?"

"I'm going to fire that assclown," Clay said, meaning Ronnie.

To prevent another strike against Ronnie, I tried to take some of the heat.

"Sir, he didn't do anything wrong. And you can't fire him for his Marines reacting to enemy contact in the manner you authorized with your operational order."

That shut him up . . . for the moment. Frustrated with Ronnie, Ron Jameson, and his FiST, he put them in charge of "babysitting" the EPWs.

Our mission in Iraq was still not complete. Late in March, we were assigned to capture Az Zubayr, a city just south of Basrah and Iraq's main port.

While driving to our battalion command post in Az Zubayr, Stan and I ran into a situation where, morally, I knew I must defy Clay's orders, though it could end my career.

I drove there to alert my battalion that our company was asking for permission, per moron Clay's orders, to mortar a "target" that my lieutenant with "eyes on" described as civilians.

We had assumed defensive positions around the port facility in Az Zubayr. Our command post would be at the base of a grain silo system with a large tower. We would keep the EPWs at the neighboring warehouse.

This battle would be tougher than in Umm Qasr, where we were consolidated and each fighting position could see the fighting position next to it, like a ring of fire pointed out. Here, because the facility was large, we didn't have enough Marines to pull that off in this new port. Instead, we had to strongpoint the defense, meaning we would not have a continuous perimeter as we had at Umm Qasr.

So we picked three strategic locations for the three platoons, with some vulnerability. Our platoons would choose the most defensible positions to the north, south, and west while using the river to the east as a natural barrier. An attached reconnaissance team would occupy the top of the grain silo in the facility, with a view of an approaching enemy.

Our weapons platoon was split up into components to augment the other three. This left the fire support team to guard the enemy prisoners of war we had collected from the action at Az Zubayr over the previous couple of days. The 3d Platoon, to the south, had eyes on the warehouses that were not part of the area we were protecting.

Stan and I had acquired a light pickup truck from Umm Qasr, and we used it to check on the defensive positions. He heard a .50-caliber rifle being fired. We stopped to listen as the reports of the rifle shots kept coming in a slow, deliberate cadence, and we couldn't tell if they were incoming or outgoing. Having been a sniper earlier in his Marine career, Stan finally determined that it must be our recon element shooting at something, because the sound was outbound.

We realized that the recon guys in the silo were shooting at something to the south. We drove to the 3d Platoon to investigate what the target was.

"What's the drama?" their platoon sergeant asked when we arrived at the platoon's position.

"That's what we're here to figure out," I said. "What's going on?"

"As far as I know, the enemy being shot at was a van of civilians looting one of the warehouses," said the platoon sergeant. As we were talking, the platoon commander, Ernie Thompson, came over to our position.

"What did you see?" I asked.

Ernie confirmed what the platoon sergeant said. The "drama" was over civilians looting.

Clay had mistakenly, I thought, called the battalion fire direction center (FDC) to clear mortar fire on that "objective." This meant that he was going to have our company drop sixty-millimeter mortars, high-explosive rounds, on those civilians. The radio call to the FDC was a formality, probably one that he engaged in after the incident with the mortars in Umm Qasr. He'd wait for the "clear to fire" response from the FDC.

I had a small window of opportunity to ensure the mortar fire mission wasn't "cleared hot," a term used by forward observers and forward air controllers to let aircraft know they can drop their bombs.

Having not seen what was going on and without time to check it out, I asked Ernie to drive with me to the battalion command post to intercede and provide the battalion with the information that the platoon commander had. I would not chance mortaring civilians, including women and children. This is where I drew the line. On my helmet I had inscribed, "No Women, No Kids," a line from the movie *Léon: The Professional*, which I'd watched ad nauseam on the way to Kuwait.

When I got to the battalion command post, I asked the battalion executive officer and the battalion fire support coordinator if Clay had just called in a "fire mission."

"Yes, but we were going to deny it because it's danger close," said the assistant battalion operations officer. "Danger close" meant that the impacts of the rounds were within about fifty meters of our forces. Because indirect fire assets have a chance of being off target, we typically do not fire them when it's possible the impacts would endanger some of our own.

The FDC had the friendly unit locations plotted and saw that a "short" round (one that has a bad charge or is defective) could hit our own men. A unit in this situation would have to acknowledge "danger close." This means that the unit requesting fire understands that the rounds might hit their position but need the fire support anyway, because they're about to be overrun.

This was not the case for us. Ernie Thompson's men might be in the impact area of the mortars, but no one was overrunning their position, and Clay hadn't mentioned "danger close."

"Good," I said. "Out of curiosity, what did he [Clay] call in for the target description?"

The battalion executive officer replied, "Ten to fourteen military-aged males dressed in black with rifles." In other words, Iraqi soldiers— Clay had incorrectly identified them or, worse, misidentified them on purpose to gain approval for the fire mission. Moron or potential murderer? I thought there was a higher likelihood of the latter.

"Goddamn. They're civilian looters. Here's the platoon commander who has eyes on that target," I said, pointing at Ernie Thompson. It turned out that the recon team was firing "warning shots" at these looters at Clay's request. This was the next step, as the Iraqis were undeterred.

Using the sniper rifle to fire warning shots bordered on idiotic. Even as a trained Marine, understanding the difference between inbound and outbound direct small arms fire can be tricky. Most military rifles shoot a projectile at supersonic speeds. This means that you'll hear the impact near you before you hear the rifle shot. Depending on the distance, the delay can be significant.

Furthermore, a tiny projectile hitting something nearby might be mistaken for other debris. Untrained civilians, like those in the van looting warehouses, may not even realize that these slow, deliberate shots were aimed in their direction.

The battalion gunner was now engaged. Having overheard what had transpired, he had some insight into what had happened earlier during the invasion regarding Clay, including, among other things, the threats

Clay had made toward one of my peers for the "fucked-up fire mission" at the border. He encouraged me to tell the battalion executive officer the full story. I eventually did, though I knew Clay would look bad.

Clay was still with the recon team in the silo. The mortar mission was denied, and it was getting dark. I brought Ernie Thompson back to his platoon, and the gunny and I made our way back to the CP.

I could not wrap my mind around this situation. I needed to know from the recon guys if they knew they were shooting at civilians. When the two of them came down from the tower that night, I asked Sergeant Donald what the fuck was going on. He did his best to explain the situation.

"So, you knew you were firing at civilians?"

"I was firing over their heads, sir," said Sergeant Mann, the other qualified sniper on the team.

"So, you were deliberately missing?"

"Yes, sir," Mann said.

"Did Captain Clay tell you to shoot to kill any of them?"

"No, sir."

"You were firing warning shots from a sniper rifle at civilians? And what if Clay had told you to engage [shoot to kill] one of them?"

"He didn't, sir."

"I don't care if he did or didn't. I need to know what kind of man you are. Would you have shot to kill a civilian?"

"No, sir!"

And that was that.

I could understand Donald and Mann's position. Because there's no law against warning shots, they couldn't argue or refuse Clay's orders, and so they shot over their heads.

Though it was determined that the targets they were firing at were civilians, distinguishing who was a combatant and who was a civilian became harder later in the conflict, as none of the enemy wore uniforms anymore. Within a month, I'd heard that Sergeant Donald had a "confirmed kill" of a military-aged male wearing a Lakers jersey, identified as a hostile target. Killing him probably earned Sergeant Donald a medal with valor.

Following the day's events, I didn't know what to expect. As I talked to the recon guys, Clay was called to the battalion command post. I hoped they'd fire him and put me in charge, although I knew the chances of that happening were slim. He was great at making himself look good and his subordinates look bad.

While taking Ernie Thompson to the command post, I figured this was the end of the line. I feared that the next day, I would be fired and sent home in disgrace.

If that happened, fine. I would not compromise my moral stance of "No Women, No Kids."

Still, I was surprised when, the next day, Clay came back and called Ronnie and me together to talk privately.

"Ronnie," Clay asked, "you know I wasn't threatening your life, right?"

"No, sir, I do not know that," Ronnie replied.

"I never pointed my rifle at you and threatened you. Did you really think that I would shoot you?"

"I guess no—I didn't think that you'd shoot me."

Though Ronnie knew that Clay wasn't a cold killer and that his comments were probably hyperbole in the moment, the threats to remove him from his position were real.

"Great, I'm glad that we've got that sorted out. You're dismissed."

Then it was just me and Clay.

"I don't know what to do with you," Clay said.

"With me?"

What the fuck? I thought for sure the battalion gunner, executive officer, and assistant operations officer had driven a nail in Clay's coffin. How could he have talked his way out of everything?

"Lieutenant Colonel Perceval is willing to move you over to H&S Company"—headquarters and service—"and have Lieutenant Hurts come over here."

"I don't understand, sir."

"Lieutenant Colonel Perceval believes that you are insubordinate, and you need to be replaced. I argued that I thought I could continue to work with you."

I couldn't believe what I was hearing. Somehow everything had become *my* fault? The actual transgressions, including the potential for firing on civilians, weren't an issue—instead, the problem for him was that I'd gone directly to the battalion command post. On top of that, he acted like he was showing me mercy!

"I think we can come to an understanding," Clay said.

"I'm not sure we can, sir. You called in mortars on civilians."

"Did you hear that fire mission over the radio?"

"No, sir, I did not. Ernie Thompson and 3d Platoon did and informed me, as I was on the road without a radio."

"Well then, it all makes sense. You didn't hear the grid that I called in." The "grid" was the location noted by the military grid reference system (MGRS).

"No, sir, I did not."

"If you had, you would've heard that it wasn't where the civilians were located, it was to the side of them."

I didn't know if that was right; it made no sense to me.

"I assumed the grid was where the looters were," I said.

"No, I was going to use the mortars as warning shots."

My jaw dropped. I was dumbfounded.

"I'm glad that we got that cleared up."

"Respectfully, sir, I don't think we have. You were going to use danger-close mortars as warning shots?"

"Yes, the rounds were going to land to the east of them near the canal."

That was total bullshit. Clay's dangerous orders had been given to reestablish his manhood, following his cowardice the first few days. He was hanging out with the recon guys and snipers, even firing a few rounds with their rifles at targets to "atone" for his combat transgressions. At one point, he said, "Killing is my business, and business is good."

"What about trading with Lieutenant Hurts?"

"I need to think about it, sir."

I left him in the room. I called for the other lieutenants to meet with me to discuss my thoughts and next actions.

After they arrived at the command post, the six of us retreated to a private area. I relayed the choice Clay had given me and explained how Clay had spun enough smoke into the situation that he wouldn't be fired and how this whole incident would be swept under the rug.

"What's best for me is to take the switch," I told them. "But I'm worried that if Lieutenant Hurts comes here, Clay will have been let off the leash, and his vengeance will rain upon all of you."

Like me, the platoon commanders couldn't believe that Clay had made up some story about using mortars as warning shots. To every reasonable infantry Marine, this was ludicrous. To scare off the Iraqis and stop them from stealing, he was effectively asking us to drop bombs near our guys, which could have resulted in some being injured or killed.

Though I knew the men could handle the vengeance, I worried someone new might be unable to intercede when Clay took his next idiotic move.

"Don't make the switch," I finally told Clay.

I stuck with my men, even though it hurt me.

"Great," he said. "Assemble the whole company. We're going to talk about this dog before we move on to the next assignment."

Sure enough, the vengeance was immediate. Some of the Marines from the company had, as a joke, spray-painted "Fox 2/1" on a stray dog they'd taken as a mascot. Further embarrassing Clay, that dog had made its way to the MEU command post and Colonel Worthington saw it. The colonel was furious that the Marines had been cruel to that animal and went right for Clay, as that action demonstrated a commander with no control over his troops. In the middle of combat, Clay berated the entire company to their faces about the dog incident.

During Clay's speech, one of the sergeants laughed.

"What the fuck are you laughing at, Sergeant Cole? You think this is funny?"

"Sir, what are the odds that that one dog walks into the CP? I don't think spray-painting the dog is funny at all—just the serendipity that the colonel saw it."

In April, I moved with BLT 2/1 to An-Nasiriyah, a historic southern Iraqi city full of residences with mud-thatched roofs or no roofs at all. The city was thought of by Iraqis as the "Garden of Eden." The Marines called it the "Wild West."

The Battle of An-Nasiriyah was fought from March 23 to April 2, 2003, between the U.S. 2d Marine Expeditionary Brigade (2 MEB), aided by the British military, and Iraqi forces.

We arrived in the city at the start of April 2003, and our battalion occupied the defenses set by 2 MEB, consolidating a lot of the area. Our company was responsible for the area fought over by two infantry battalions.

This was happening after our battalion had seized the western half of the city as part of the mission to rescue Private First Class Jessica Lynch. A teacher, actress, and U.S. soldier, Lynch was captured by the Iraqis and seriously injured on March 23 when her supply unit was ambushed. U.S. special operations forces (SOF) rescued her on April 1. It hit all headlines, as this was the first successful rescue of an American prisoner of war since World War II, and the first ever of a woman.

From the power facility between the Euphrates and the northern canal, we patrolled the area to find unexploded munitions or remaining enemy combatants and to block any enemy reinforcements from the north.

Within a short time, the other infantry companies that had conducted the attacks on the western side also consolidated, and now our MEU was covering the area that the MEB (about three times the MEU's size) had during the initial battle.

Our company moved from the northeastern section, an area known as "ambush alley," to the southwestern section of the city. We occupied a building used as an antiaircraft position by the Iraqis. This was evident from the antiaircraft guns in the yard and the room in the building that resembled air traffic control. A red crescent was painted on top of the building, which theoretically meant it was a medical clinic and therefore off-limits from our strikes.

From that position, we spent the next couple of weeks patrolling the whole area, slowly blacking out the respective zones on a large map we kept in the company command post. Outside of a few false reports of Ba'ath party members operating in the area and misunderstandings with some locals, there was no more action. By late April 2003, all the hostiles seemed rounded up.

Other Marine units had marched to Baghdad, Iraq's capital. Saddam Hussein was nowhere to be found. By April 9, after a six-day battle, U.S. forces captured Baghdad. By May 1, President George W. Bush declared that the first phase of the war was complete.

We, the Marines, had kicked in the door. With our job completed as an expeditionary force in charge of invading Iraq, it was now time for the Marines to go home, while the standing Army soldiers cleaned up the rest of the country.

That was it. No more combat. I had barely been in the thick of it. Was the deployment even worth it? I went to Iraq and never fired a rifle. I wasn't a hero; I felt like an imposter.

CHAPTER 6

MEDALS

Harvey pulled the bandage back and showed everyone his wound...
"Purple fucking Heart, bitches. You know how much pussy I'm
gonna get back home!"

—Phil Clay, *Redeployment*

From Bahrain to San Diego, with stops in Dubai, Australia, and Hawaii, sailing home from Iraq took us about a month. These port calls gave me time to venture into the town on liberty, to decompress, walk on dry land, eat local food, and, most importantly, ponder what had transpired.

Many thoughts streamed through my head, mostly negative. Having seen little combat after three deployments, with little chance for deployment in the future, I thought I had accomplished nothing.

The ambush on the way to Umm Qasr didn't count. Despite the danger and near-death experience, I hardly saw a "bad motherfucking hajji," other than a small platoon of about thirty Iraqi soldiers who had surrendered.

Hell. Since the enemy was invisible, I hadn't even shot my pistol, let alone an AK-47 that I'd picked up or Clay's rifle that he irresponsibly forgot and left behind. All I did, I thought, was seek cover.

If push came to shove, could I kill? I didn't know. Maybe not. I'm a peaceful kind of guy. I heard that in World War II, when Marines hit the beaches, many didn't fire their weapons, even when staring the enemy in the face.

Either way, the ambush was hardly combat that could earn me the coveted CAR.

To get a CAR—that tiny piece of ribbon pinned to our uniform that tells the world we went to war, something Marines would march to hell and back for—requires engaging in combat. That meant shooting at bad guys, which I had not. My one and only chance, and I missed it!

Furthermore, having challenged my commanders before the invasion with regard to training and challenged Clay during the invasion, I would receive *no medal at all*. Even worse, neither would my platoon commanders. On the ship home, vindictive Clay disapproved of the fifteen to twenty combat awards that my other lieutenants and I had written for our men.

I argued to the higher-ups that every single one of the E-5s in my company deserved an achievement medal with "V," a device added to an award that indicates an achievement made during combat. They'd successfully led Marines in combat, so the "V" was necessary to show it was earned on the battlefield.

Somewhere between Bahrain and Australia, the arguments concerning our Marines being recognized with a Combat "V" came to a head. I pointed out to Clay that he had the opportunity to disagree with the recommendation but that he could not outright make the lieutenants rewrite their nominations.

"Sir, there's a place on the form where you can disapprove of the 'with combat distinguishing device' as you pass it up the chain of command."

"It doesn't work that way."

The fuck it doesn't, I thought. "Sir, that's the way the form is designed."

"It has to be consistent on the form."

This meant that Clay and the battalion commander had to agree in order to get the addition of the "V."

"What is the battalion commander's guidance, then, regarding these devices? Dean has researched the awards manual, and this is our conclusion—that our guys rate them."

Later I'd be forwarded an email from Clay regarding the battalion commander's guidance. Clay had mistakenly forwarded the battalion commander's response to him regarding the "Vs" without editing the previous back and forth. Not quite the classic "reply all," but it still turned out to be telling.

Scrolling through the messages, I saw that Clay had started the conversation with the following: "Sir, I apologize for the lateness of these awards nominations. My lieutenants need to take the MCI [Marine Corps Instruction] for writing. These awards read like they wrote them with crayons."

I took the "reply all" off and responded directly to Clay, saying that he might want to address the comments that he made, as now the lieutenants could see what he had said about them.

"They shouldn't read that. It's the private communication between me and the battalion commander."

What a moron!

"Sir, you forwarded it to them; there's no way not to read it."

That opened a bigger schism between Clay and me. On our smaller ship, this new conflict meant that he would keep to his stateroom to avoid me and the lieutenants. Where previously he might eat with his men, now he wouldn't. He wasn't checking on his Marines and wouldn't talk to me in person.

Once we reached Australia, I was told that I would be moved to another ship, where the command element was to "work" on these awards. That was fine with me; I thought that being on "the big deck" meant that I could better advocate for the Marines.

I was wrong. The majority of the awards were either denied or recycled and approved without the device. In the end, I think that only one of the staff NCOs—and maybe a couple other Marines who had used the mortar tube in the handheld mode to fire on the enemy—would receive "heroic" awards.

At this point, knowing there had to be a witness of heroic action, I realized that none of the lieutenants in our company would be recognized or awarded. It didn't matter that they led the Marines in the port clearing or that they had reacted to the indirect fire and led patrols in An-Nasiriyah. They'd get nothing for now.

Though the four lieutenants would likely make another deployment with the battalion, which gave them another chance to get an end-of-tour award, this *was* the end of my tour. I would get nothing.

I brought this to the attention of my battalion executive officer. He flaked on me by saying he couldn't recommend me or any of the other lieutenants for actions in Iraq because he hadn't personally witnessed them. *What a crock of shit!*

If you have been in combat, not getting a medal is a disgrace. From the Purple Heart, created by George Washington to recognize the combat wounded, to the Medal of Honor, given to combat men and women who had gone above and beyond the call of duty and usually sacrificed their lives for others, medals play a huge role in the extrinsic reinforcement of the military.

I did get some non-combat medals. While we were approaching Kuwait, my unit learned we had earned a medal for Operation Southern Watch—the operation started after the first Gulf War in 1991 to ensure a no-fly zone in the northern and southern parts of Iraq—and I got a "pump," or deployment medal (been there, done that).

Also, I earned a prestigious ribbon, the Presidential Unit Citation. This one has been infrequently awarded only twice, I believe, since the start of the Iraq War. Members of the military who have this citation ribbon likely earned it from being part of the invasion in 2003.

While these medals and the ribbon distinguished me from the rank and file, they meant little in terms of military prestige, and even less the further I got in my career and the closer to the top-tier units. These weren't combat medals. The medals and awards for heroism require the enemy to participate.

Having no combat medals would be a curse. My peers would view me as inferior. Since future promotions and selections would be tied to having gotten a medal, my progression would be halted. Even worse, I would have no choice but to continue to work for asshole commanders in the future.

I felt like a failure. I thought about throwing in the helmet.

To make matters worse, Clay would be promoted and given an award with a valor device, a medal with the capital letter "V."

This infuriated all of us. That moron had put our lives in danger. If some of the details about his antics got out—like forgetting his rifle when he ran for cover, which I later recovered, or having to be coaxed into participating in the Umm Qasr port clearance operation—there was no way that he could justify his valor device.

There were more transgressions. That night after the skirmish outside of Umm Qasr, I wrote in my journal of my close call with death. Later, Clay grabbed it, and, looking to transform my personal journal into a distorted historical documentation of events, slashed my commentary to improve his image. He blocked out where he had been in a state of shock. And he edited his tactical mistakes and early inaction. I wanted to punch the guy.

He handed the journal back to me and said nothing about his changes. I added nothing else. Let someone else be this jerk's personal historian.

Furthermore, I witnessed him "cheating" to get a medal. Once the combat died down in April and our time in Iraq was coming to an end, Clay chose to be a part of the breach into a suspected enemy compound. As a company commander, this was not his role, but rather a brash and foolish attempt to get into the action, as the following demonstrates.

We had received some information that one of the Ba'ath party leaders may still be inside this building. The information fit with some of the intelligence we received before entering the city, indicating that the Ba'ath party leader's residence was in that location.

1st Platoon had cordoned off the house and was preparing to enter it.

I was out with Stan checking on the positions when 1st Platoon got this information from an English-speaking local. We radioed in that we were going to make entry. Clay radioed back to wait for him. I figured he wanted to be there if there was a chance to shoot at someone in order to bolster his image. So far, the only shooting Clay had done was at a lock at the port facility of Umm Qasr, where he had struggled to destroy the lock in order to get his vehicle through the gate. Unsurprisingly, he missed the lock and had to shoot a second time before the other Marines pointed out a way to bypass the gate.

The 1st Platoon commander and I waited for Clay near the compound. When he arrived, he chose Robert Shore to breach the door. Clay would then make entry. This move was risky and unnecessary; the first person in is vulnerable to everything in the room as he crosses the threshold.

Shore blew the door off its hinges with a shotgun, and Clay went in.

This is it, I thought. *That idiot is going to get himself killed.*

I was unsure if I felt bad about it. After all, this was the guy who had wanted me replaced not three weeks earlier.

There was no one inside, so we locked the building back up as best we could to prevent any looting and went back about our business.

Clay's act of being the first one in smacked of searching for the Silver Star. He wasn't alone among commanders. I heard that our battalion commander, the sergeant major, operations officer, and their radio operator were doing similar things. That particular "fire team," or group of four Marines, seemed to be patrolling around the whole area all the time looking to insert themselves into some kind of action. They had to be close to someone shooting at bad guys to justify the awards they were writing for themselves in their heads.

Despite not having done anything heroic, Clay, by dint of being a commander, would likely receive the Bronze Star, especially as our battalion commander had received a Bronze Star *and* a Legion of Merit as an end-of-tour award. Our battalion executive officer would receive a

Bronze Star with "V," even though all he did was ride in a vehicle while the Force Recon Marines attacked buildings with no bad guys in them. The battalion sergeant major would receive one for "rallying the troops" during an ambush on the LAR company attached to us. Of course, he did the rallying several miles away from the action, in the back of an armored vehicle.

On their face, the citations were ridiculous. If the battalion commander recommended the citation, it was rubber-stamped and approved.

The Marines of Company F, 2d Battalion, 1st Marine Regiment (F/2/1) were irate at Clay getting a prestigious medal. Even before we landed in Kuwait, the Marines wanted to submit a letter of no confidence in Clay because of his actions during training.

To them, Clay was our generation's Captain Sobel, a harsh company commander during World War II who constantly berated his soldiers, punished them for minor infractions, and frequently canceled weekend passes. Sobel had been recently portrayed in the widely popular *Band of Brothers* series recently aired on HBO. So disliked was Sobel by the soldiers that the NCOs wrote and signed a letter of no confidence. The letter was delivered to the battalion command, and Captain Sobel was reassigned.

Thinking the same technique would work with Clay, the Marines F/2/1 updated Clay's actions in Iraq in that original letter of no confidence. When we landed at Camp Pendleton in July, they delivered it to the 1st Marine Regiment. That started a new shit show.

Before that happened, onboard a ship sailing from the States to Kuwait, before the war, one of the senior enlisted Marines approached me about delivering that letter.

"We've got just about every squad leader and NCO signing the letter; that's fifty-two or so signatures. The only guys who are worried about signing are the career guys."

The letter highlighted the times Clay had lied to the battalion commander, as well as other transgressions that I wasn't aware of.

"What do you think, sir?" he asked me. "Would it work, or will it do nothing?"

Knowing Marone had never been relieved of command despite worse transgressions—not going to the field, leaving a Marine behind at the beach, drastically failing at tactical exercises, etc.—I was skeptical.

"Not a fuckin' thing will happen with the letter. There's no way, now that we're sailing, that they'd replace him."

Knowing Clay would spit fire if they sent it, I suggested that they shelve it.

Nevertheless, when we got back to Pendleton on a Thursday in July 2003, the updated letter was turned in to the regiment the next day. I was retrieving my gear from the ship in San Diego that Friday and had nothing to do with how that letter ended up in the hands of the regiment, nor did I know where it was delivered.

But on Monday, I was called into the office. They told me I was to be "soft-relieved" or "fired" as the executive officer without permanent marks in my record.

Robert Shore was going to assume the duties of executive officer. I was told I could either take a vacation until it was time for me to execute my next set of orders in Norfolk, Virginia, for service with the Marine Corps Security Force Battalion, or I could work for the month of August in the operations shop.

In my defense, I tried to argue with the battalion executive officer.

"Sir, wouldn't it make sense to leave me in place for the incoming company commander to use my expertise as he is navigating the company?"

"We think it's better to just move you now."

This pissed me off.

"Sir, Clay only has five days left of command." He was going to relinquish the command of the company to the new captain that Friday. "So, there shouldn't be a personality conflict issue anymore."

"We think this is better. It's a clean slate."

I was steaming. "Sir, this wouldn't benefit the Marines," I muttered calmly, struggling to not lose control. In the military, there's an unspoken rule that you don't show anger to your superiors.

"Shore will be fine as the XO [executive officer], and this way you can move on."

Move on?! Where? The sense of injustice consumed every cell in my body.

I saluted him, turned on my heels, and stomped away.

Meanwhile Clay would still march as the company commander in the battalion commander's change of command. There would be no ceremony for the incoming commander taking over the company from Clay. Promoted to major, Clay would move to the regiment and then leave active duty for the reserves.

Later, I learned that Clay had accused me of delivering the letter to the regiment, among other insubordinate acts. There was no way I could vindicate myself. On top of no combat and no medal, I now had no respect, no appreciation, and false accusations leveled against me. The only thing that saved me was a promotion to captain in September 2003.

I started working at a fleet antiterrorism security team (FAST) company in the security forces. FAST platoons are an elite unit of the U.S. Marine Corps that are capable of rapidly deploying to immediately improve security at U.S. installations worldwide. They are also capable of deploying as an infantry quick reaction force.

Shortly after working at FAST, I got a call from the new battalion commander of 2/1. He apologized for what Clay had done. Clay had allegedly forged the battalion commander's signature on fitness reports, which are officer evaluation documents. I got a copy from the first sergeant there at F/2/1. Mine read, in part:

> Rated 6 of 6 officers in company. He was counseled, taught, guided, and mentored during the reporting period. Despite my best efforts, he continued to only meet the minimum performance standards. This period covers time of personnel turnover with his Reporting Senior [RS] and Reviewing Officer [new battalion commander]. Performance was marked

with shortfalls in judgment, loyalty, and decision-making ability. Upon return from deployment, Marine Reported On [me] has joined the [Battalion's] Operations Section.

There were dated entries throughout the report. For example, on July 3, it stated that I had been "counseled for exercising poor judgment as he formed and represented the leadership of the Company without informing the Company Commander."

I couldn't argue with this one. The Marines of my company wanted to present the captain of the ship that we sailed over on—a Navy commander we admired who looked like a George W. Bush doppelgänger—with a large Iraqi flag they had all signed. Not wanting Clay there to taint the gift, they asked me to pass a message to the ship's captain to meet them on the flight deck for the presentation.

There was no formation. I didn't represent anything. All that happened was that the senior enlisted Marines unfurled the flag for him. The ship's captain made some comments. I didn't even hear those comments, because the platoon commanders and I were watching from a deck above and weren't part of the event.

Another entry dated July 10 said that I had been "counseled on the reluctance to except [sic] my intent regarding combat awards."

Bruce Dean and I had researched the awards manual to ensure that our Marines were recognized for their actions during the invasion. We'd insisted that those Marines awards carry the combat "V." Clay rejected the inclusion of the device. He said their actions didn't count as combat and weren't worthy of valor.

By this point, Clay refused to talk to us face to face. He would sit in his stateroom next to the one occupied by the five lieutenants and write emails to them. I'd knock on his door, hoping to talk things out, and he wouldn't answer. So, I emailed him instead, directing him to the box where he could nonconcur with the "V." He would have to, because he couldn't simply force the lieutenants to change their recommendations.

Given the impasse, it was decided that I should move to the big deck, the larger ship in the MEU, where the battalion command element was working to facilitate award approvals. I submitted them with the combat device, and only a few were approved.

Yet another entry in the report asserted that on July 16, I had "contributed, endorsed, and delivered a letter with the expectation to tarnish the reputation of the chain of command without first attempting to provide accurate and proper professional corrective action to resolve disputed issues."

As I have noted already, and as Clay probably refused to believe, I tried to shut that letter down twice—once before the invasion and then again sailing home. I have no idea who had the original or if copies were made, or whichever version ended up in the hands of the regiment.

Was I upset that it became public? No. Clay fixed all his hatred on me. In his mind, I was the only person who could have done this. He probably thought I forged all the signatures, too.

I checked with the other five lieutenants—Robert Shore (1st Platoon), Bruce Dean (2d Platoon), Ernie Thompson (3d Platoon), Ronnie March (Weapons), and Ron Jameson (Artillery Forward Observer). They had all received poor reports from Clay.

In his apology to me, the new battalion commander promised that Clay would be court-martialed for forging his signature. By this point, I'd moved on with FAST, and as long as they pulled that report from my record, I didn't care what happened.

All's well that ends well, and eventually our company earned the CAR. As evidence that I was fired upon by the enemy, I had fortuitously taken a picture of myself with the shrapnel in my notebook. That artillery fire evidence turned out to be enough for us to be awarded the CAR, even though we weren't able to fire back.

At FAST, I was assigned to a platoon scheduled for a Middle East deployment. I was working with a higher caliber of Marines, most of whom had missed the invasion. If HQMC pulled that egregious report, my deployment to Iraq probably wouldn't hold much weight. I would have an "admin filler" in lieu of a combat fitness report.

Though there was nothing that I could do to change what Clay had written, I was willing to testify that he had fraudulently signed the document. Clay had claimed he provided me with a signed copy of the report, which he had not.

Clay's legal process was drawn out. Eventually I was deployed to Baghdad with FAST in November 2003. While there, I sat by the phone one day at 2:00 a.m. to be called in remotely as a witness. I wasn't called.

Later, I discovered the charges against Clay had been dismissed. The legal aide informed me that Clay's lawyers did a better job of arguing that the signatures were valid, and because the prosecutor ended up getting deployed to Iraq, the government was never able to make a strong enough case. This upset the new battalion commander, but there was nothing he could do.

I was pissed that I'd woken up in the middle of the night to testify from Baghdad and didn't get called. I didn't even get a courtesy call to tell me what had happened. However, I probably didn't feel as angry as the new battalion commander, who was basically called a liar by the defense.

As for me, prior to hearing the case's outcome, I had taken some solace in knowing that this probably meant the end of the road for Clay. Now I knew it wasn't, and that was the end of that shit show for me. It was time for me to move on.

PART II

WAR AND FAMILY

.

CHAPTER 7

DATING THROUGH EMAILS

Roses are red, cammies are green, I'm in love with a U.S. Marine.

—Common saying among partners
of U.S. Marines

In April 2004, I was stationed with FAST in Norfolk. Two Marines I knew there were dating these local gals. They had a friend they thought I might like.

"She's a real looker," one of them said. "Let's all meet for dinner."

Why not? While meeting women was not foremost on my mind—my only desire being to get back to Iraq and combat—I could still use a lay. I disliked getting sex from one-nighters after a drunken night at a bar, so this at least seemed preferable.

The minute I strolled into the bar and grill, my eyes were immediately riveted to a tall and willowy gal with long, chestnut, wavy hair and dark almond eyes.

"Varpas, meet Amy," my buddy said.

She held out her hand and flashed a broad, white-toothed smile. "Nice to meet you, soldier boy." She patted the bar chair next to her. "Have a seat."

I couldn't take my eyes off this bubbly lady, two years younger than my twenty-six years. We conversed easily, though she did most of the talking. She was an engineer. I was impressed. My mom was a biochemist, and I always liked sharp, smart women.

I thought we had hit it off. Then she announced, "You seem a great guy. But I'm not keen on dating a Marine who would be away on deployments. And . . . I would never marry a Marine."

Well, I thought, *there goes that.*

She said goodbye with a "no hips touching" hug, the same hug she'd just given my buddy. My stomach twisted in disappointment.

Before turning to join her girlfriends, she gazed into my eyes, lingering there a moment. Should I ask for her phone number? No. If she had been interested, she would have given me a real hug. Still, thoughts of her lingered in my mind.

Meanwhile, in May 2004, as a captain with an O-3 pay grade, I was sent with FAST to Bahrain to assist the Navy in making assessments on the security in Umm Qasr and along the oil pipelines. After completing the assessment, the team and I headed north to Basrah and then to Baghdad, where the top guys of the team briefed the coalition provisional authority.

My FAST buddies who'd rotated in to replace me and my platoon from our deployment from November 2003 to February 2004 were still in Baghdad. While searching for them at the Al-Rasheed palace, I ran into those two Marines from that night with Amy.

"Whatever happened to Amy?" I asked.

"Amy? Didn't you get her phone number?"

"No."

"What the fuck's wrong with you, dude? You two hit it off."

"Shit. I wanted to."

They laughed. "What a dick."

"Hey," one of them said. "I'll reach out to my girlfriend. Hopefully Amy hasn't hooked up with a hottie Marine."

A week later he sent me an email with Amy's email, as well as the following message: "Get some!" He said he gave her *my* email, as well. Should I write first? Would she answer if I did? She might not even remember me.

I put it off. Later in the month, when I'd returned to the FAST platoon in Bahrain, I got an email from Amy: "Hey, how are you? Always wondered why you never asked for my phone number. I would have given it to you."

We emailed back and forth for hours that night, as well as the next night and the next. "Nice to get your emails after sweating bullets in the desert all day," I wrote.

"Happy to help the Marines. I'll write every day. Promise."

"Why would you? Is it my looks (LOL)?"

"Yeah. Something about those baby blues and winning smile."

My confidence puffed up. Still, I was skeptical. "You will for the first week," I wrote, "and then every other day and eventually maybe once a week."

I was on my fifth deployment. Relationships I had tried to build had gone that way. Long distance never seemed to work. It didn't help that in four years, between those five deployments, I had moved from Quantico, Virginia, to California, then twice within California, then back to Norfolk, and then once more within Virginia. I just wasn't around. Why make an effort to keep in contact with me?

Now, the woman who had said she wasn't keen on dating a Marine took the challenge and emailed me every day—all in all, over 852 emails, with occasional instant-messaging chats. Some were long emails, in which she'd write the story of her life, and others were short, describing shenanigans at work; it was how we got to know each other.

I plugged into the internet daily. Then I left Bahrain for Saudi Arabia to reinforce the embassy there, followed by a stint at Camp Lemonier in Djibouti, in the Horn of Africa, where there wasn't always a computer terminal, much less the internet, for me to match the story of my life with hers. Still, when I finally got online, I saw that she had kept her promise and written daily. Our email communication lasted from June until the next time I would see her in November 2004.

Amy had moved to California for a new job. I was still assigned to FAST in Norfolk. I asked if she would come to Norfolk to welcome me home.

"I'll hop on the next plane," she wrote.

I beamed. This would be our first real time together after that brief night almost six months earlier. Would this work out? We hardly knew each other.

On my way to the luggage pickup, I spotted her, dressed in a sexy, tight, red pencil skirt, her black hair long and flowing. She waved and flew through the crowd into my arms. She hugged me fiercely, and we kissed passionately.

"Missed you," she said.

My body surged with excitement. What a homecoming! No one had ever been there before to greet me coming home from a deployment.

We spent a delightful week together. Sex was great. Not having been a one-nighter guy, I hadn't rolled in the hay for a while.

By the time she returned to her job in California at the end of the week, we knew we had magic.

I invited her to spend Christmas and New Year's Eve with my mom, my brothers, and me in New Bedford, Massachusetts. (My parents had divorced, and my father was living in Portugal.) That was the soonest I would have furlough.

"Love to meet your family," she said.

My mother welcomed her like a member of the family.

Amy and I locked tight for a week, hardly leaving each other's side. In January, I returned to Camp Lejeune, North Carolina, where I spent a month getting ready to deploy to Iraq for a year.

Amy visited me there. There was little to do in Jacksonville and certainly no Michelin-rated restaurants. She felt a little foolish in a cocktail dress as we headed to an Applebee's for Valentine's Day.

I had been teasing her with a riddle to guess her Valentine's gift. "It's small, red, and close to my heart."

She ventured many guesses but never got it. That evening, I took her hand and placed a special dog tag in it. Most dog tags are silver. To help identify and process casualties, they have the service member's name, blood type, identification number, and religious preference.

I wanted her to have it because if something happened to me, she wouldn't find out for some time, as she was not next of kin. Those who have medical concerns, like an allergic reaction to morphine, get an additional red dog tag as a military medical alert bracelet. I had a red dog tag made with her name and phone number on it so she would be called if something happened to me.

She shuddered when she saw what it was. Though she appreciated how I had gone out of my way to have it made, the reality that I could die flew in her face.

"I didn't mean to frighten you," I said.

"I live with that fright any time you deploy."

I squeezed her hand. What could I say? She was officially the girlfriend of a Marine warrior.

Parting was painful. I hugged her tightly before she boarded her plane back to California. She gushed about how she loved my muscular, lean body and would miss waking up to "your handsome face." I stroked her hair.

"Stay safe," she said and gave me one last tight embrace.

Amy kept abreast of any news regarding Iraq. Even though, by this time, the myth of weapons of mass destruction had evaporated and Saddam had been dragged from his hole, the newspapers were replete with stories of rampant violence despite his capture. Throughout the winter, she read about deadly bombings across the country and feared for my safety.

"Don't worry," I assured her. "I'm doing the Special Forces mission of advising, assisting, and training the new Iraqi Army so they themselves can secure the country. It's a relatively safe assignment."

Of course, I didn't tell her that my ultimate goal was to return to combat.

The security forces battalion commander wanted to rotate more captains into the platoons, and I asked to be an individual augment to

2d Marine Division (2 MARDIV), as this was my best chance to get back to fighting.

In February, I joined them in Al Anbar Province, Iraq, my sixth deployment. In July 2005, I switched from being a staff officer to the job I volunteered for, a military transition team (MiTT) member with the Iraqi 7th Infantry Division.

As a MiTT member, I would be one of a few Americans with an Iraqi unit in Al Anbar Province. There would be twelve of us Marines and about two hundred Iraqi soldiers. "Advise and assist" meant that we would go outside the wire with them and patrol.

This was riskier than patrolling with an all-American unit. The enemy—at the time, mostly al-Qaeda in Iraq (AQIZ)—would avoid units of Marines patrolling. A softer target, however, would be an Iraqi Army unit. The enemy would see an Iraqi unit patrolling and choose to attack them with more frequency. This meant that MiTT members would be in the mix by being in proximity to these Iraqi forces.

In almost every conflict, Marine units are quickly identified and avoided because of the ferociousness of their attacks. In World War I, the Marine units in France earned the nickname "devil dogs" from the Germans because of their relentless pursuit of them. In Korea, the Marines, wearing tan gaiters on their boots, were called the "yellow legs"—the North Koreans and Chinese avoided these Marines and attacked U.S. Army units instead. U.S. Army Major General Frank E. Lowe wrote, "The safest place in Korea was right behind a platoon of Marines. Lord, how they could fight!"

In later conflicts, Marines would be known as "white sleeves," because they rolled their sleeves differently than Army units. By 2005, after the events in Fallujah and Ramadi, the Marines had again established their reputation as a "hard target."

The AQIZ forces knew the Iraqis had little will to fight, as most were stationed far from their homes; my Iraqis were fighting in Anbar, a day's drive from their homes in the southern part of the country in Numaniyah or Diwaniyah. They were mostly Shi'a Iraqis fighting in the Sunni Triangle.

This would be the softest target for AQIZ to hit. And indeed, they did in many places like Rutbah, Rawah, Al Qaim, and Baghdad.

I hid this danger from Amy. My time in Iraq in late 2003 and into 2004, I told her, was "hardly combat." I was in the Green Zone, the heavily guarded diplomatic and government area of closed-off streets in central Baghdad where U.S. occupation authorities lived and worked. We were staying in the Al-Rasheed palace on the Tigris, which had several massive "Saddam Head" busts on the roof. Later, this palace became the U.S. Embassy.

Nearby was the Al-Rasheed Hotel, where we consulted on force protection measures—meaning, we were called in to see how we could improve the defensive positions, patrolling efforts, and antiterrorism posture of the hotel and palace. We were asked how we could use our FAST Marines to prevent subsequent rocket strikes and attacks on the Green Zone.

I emailed Amy about how, after the Gulf War ended, a tile mosaic depicting President George H. W. Bush was installed on the floor of the hotel lobby. This would force visitors to walk over his face to enter the hotel, a major insult in Arab culture. After the invasion in 2003, U.S. soldiers smashed the mosaic and left a portrait of Saddam behind.

"I walked over Saddam's face."

"LOL," she emailed back.

During my time at the palace, they lowered the "Saddam Heads" off the roof. I took a picture of my corpsman peeing on one of the busts, next to a sign that said, of course, "No Peeing," and I emailed a copy to Amy.

"What was it like staying in the Palace?" she asked in an email.

"Like being in the lap of luxury. I have a bed and a mattress, catered food, a gym, hot showers, and access to computers and phones, so we don't have to lose touch. It even has a pool. It feels more like Club Med than Club Combat. The other night I met some NFL cheerleaders and wrestling personalities. And David Letterman."

"Don't get too comfy."

"No way. There's a Muslim call for prayers five times a day starting at dawn. It echoes throughout the city and is irritating."

One night, years later, while on base in California, I awoke in panic hearing that prayer. It turned out to be a drill for the Marines. I had to pinch myself to be certain I wasn't back in Iraq. After I retired, as a prank, I sent some of my Marines buddies an alarm clock with the prayer on repeat.

For the most part, we felt safe in the palace, because our command posts were in partly fortified buildings with a large amount of concertina wire to keep out the enemy.

For a while, body armor was not even required when going to the nearby souq because the bazaar was in the Green Zone. So, too, was the combat surgical hospital that was still receiving casualties from various places in the country, including right outside the Green Zone. Because of massive security and loads of checkpoints, the entire Green Zone was considered to be inside the wire. Only those on guard duty had to have their kit on. Everyone else could take a handheld radio, get into a jeep or Suburban, and more casually drive to another point in the area.

Still, despite the apparent safety, we knew it was tenuous. The palace was the head of all coalition forces, making it a symbolic target if the budding insurgency could find the will or strength to strike it. The enemy's most dangerous course of action would be an attack from the south, coming over the Tigris and hitting the palace from the cover and concealment of the groves between the palace and the river. As it turned out, the palace did receive threats that we mitigated as an antiterrorism team, but we prepared for the worst.

Once, the palace *was* hit with indirect fire. No big deal. It was hardly *combat*. When rockets had hit the Al Rasheed hotel, the damage done was significant nonetheless. Some of the rockets had destroyed rooms and set fire to a couple of the floors. There were some cracks in some of the stone walls, though they were imperceptible if you didn't know where to look for the rocket impacts.

Another year would go by before I would see Amy again, another year of dating through correspondence. Somehow it worked, in part because Amy was a biomedical and electrical engineer working on microchips, which kept her extremely busy. She didn't need a man to define her.

Far away from any palace, from 2005 to 2006, I lived just outside of the sprawling Al-Asad Airbase with a battalion of Iraqi soldiers and eleven other U.S. Marines acting as a military transition team. The Al-Asad Airbase was the second largest U.S. airbase in Iraq. The Marines there had no frame of reference that they were even in Iraq, so far removed were they from the wire. The airbase had a "salsa night" and an intramural softball league. The Marines from the air wing on the base wore civilian attire to their dances and ate at the newly flown-in Pizza Hut and Burger King trailers.

Because so many Iraqi soldiers needed training and the standard special forces couldn't manage all of it, we Marines were there primarily to help the Iraqis reconstruct their lives, not necessarily fight battles. I wrote about this to Amy. My mission was patrolling with them and advising and assisting them in their combat operations, the role I told Amy I had originally been assigned when leaving for Iraq in 2005. See, Amy? Piece of cake here. No need to worry.

How far from the truth. We were, at most, only nine Americans on a patrol or combat operation with up to two hundred Iraqis, looking for IEDs. Even when things seemed peaceful, and while most patrols or troop movements were uneventful, peril floated in the air. Ventures outside the wire were fraught with danger. At all times, some part of me prepared for when the enemy could strike. Even on my last combat deployment to Iraq with the Marines—the one with Company C from 2006 to 2007—I kept my head down so it wouldn't get shot off.

Living in this unstable reality, you live in terror, knowing you can die at any second and that most everyone in this country hates you. No one escapes this feeling of dread. I did not share these fears with Amy, though; she was well aware of the reality of being in this hostile territory.

Before I left for Iraq, at her suggestion, I made out my will and took out a life insurance policy.

Amy knew how Clay had prevented my crew and me from getting medals. She was overjoyed when much later, in late 2007, I wrote her, "Finally, I'm being awarded a medal 'with heroic device' for those first couple of months of 2006 in Iraq. And I'm a runner-up for the Leftwich Award for outstanding leadership across the Marine Corps."

Still, these awards weren't enough for me. I had not received them because of brilliant initiative or because I shot bad guys. My "bravery" happened because I simply did what I was told to do. I went to the places my command said to go. I relied on the equipment I was given. I executed the IA drills as I knew them and made what seemed like simple command decisions under fire. This was not heroism. It was following orders. This was not the warrior I wanted to be.

Back in November 2005, I spent a two-week furlough at Amy's apartment in Orange County. When I got off the plane, she flew into my arms. We kissed madly.

Seeing her was thrilling. We had a whole two weeks together. Yet after two days together, I played video games and watched movies non-stop to fast-forward through the furlough, itching to get back to the desert. What was wrong with me? At the time, I had no idea I had the symptoms of WWS—the desire to return to combat.

Though I didn't discuss this desire with Amy, she sensed it. Most women would have nagged me to get up and go out with her. Not Amy. She took my rude behavior in stride. How lucky I was to find someone as understanding as Amy!

When I got my plane ticket, my flight from California to Texas would not be in time for the rotator flight to Kuwait. I ended up in a hotel in Texas for the evening and took the rotator the next day. I called Amy and told her I was spending the night in Texas. She hopped on a plane and met me within hours—she was the one.

CHAPTER 8

FAMILY LIFE AND DEPLOYMENT

We are in this together, no matter how far apart.

—Common saying among U.S. military families

In February 2006, I headed back to Camp Lejeune after my MiTT deployment. Amy was there when my plane landed.

Soon after, I returned to FAST in Norfolk. Amy hopped on a plane from California and met me there. In March, we decided to get married, but we postponed the wedding ceremony. With what little money I had, I bought her an engagement ring, a fake diamond, with the promise that once I'd saved up enough money, I would get her the real thing. Amy, always "real" herself, thought my stand-in ring was charming.

In September 2006, I reported to Fort Benning, Georgia, for the maneuver captain's career course (MCCC). This course would take six months to complete. That meant another six months would go by before we could marry and live together.

At Christmas, I had a furlough that I spent with Amy and her family. It was the first time I met her parents. She was flying in from California, and I was flying in from Georgia.

As luck would have it, the airline lost my luggage, and Amy's flight was delayed. I met her parents at midnight, without her, and borrowed her younger brother's clothes. Amy arrived the next day, and that one evening of awkwardness was over.

After the holiday, Amy returned to California, and I went back to Fort Benning and the MCCC. The two of us clung together fiercely at the airport gate. "Parting," Amy consoled me, "makes each reunion more exciting and sexy."

My best friend as well as my lover, she helped me study and prepare assignments over the phone or through email.

While in Georgia, I lived in an apartment with an inflatable bed, a large leather chair, a large TV, and little else. While that was enough for me, Amy thought I should have more furniture and bought me the matching ottoman to my leather chair. Once we got our own home, she would be our decorator, another of her many talents.

"It's stinky here," I emailed Amy about the apartment. "The previous tenants smoked a lot, and the place reeks of cigarettes. The management used smoke bombs, but it still stinks."

She suggested I use charcoal to soak up the odors. I did—each corner in my barren living room had a paper plate with charcoal briquettes. Still, it took six months until there was no smoke smell.

I graduated from MCCC in February 2007. I was at the top of the class and presumably had my choice of assignments. I was assigned to 1st Battalion, 1st Marines (1/1) as a company commander. I was ecstatic. To congratulate me, Amy sent me a teddy bear.

In March, I attended and graduated from Basic Airborne School ("Jump School") in Fort Benning, Georgia. Afterward, I reported to 1/1 at Camp Pendleton, California, where I would be close to Amy—finally. The oldest, largest, and most decorated Marine division, the 1st Marine Division is a force of twenty-two thousand men and women organized to conduct combat operations, and I picked up company command almost immediately.

On March 31, 2007, Amy and I officially walked down the tiny aisle of the Ranch House Chapel at Camp Pendleton. My battalion chaplain married us in front of a small group of friends and family. We had our reception at a small restaurant by the ocean and our first dance on the pier, a romantic beginning.

I had a few training evolutions left, including a month in the California desert at the Marine Corps Air Ground Combat Center in Twentynine Palms, doing a "Mojave Viper" exercise. During training, Amy fetched me from the desert to take a passport photo at the post office and apply for a new passport. I needed one for our honeymoon in Bora Bora, an island in French Polynesia northwest of Tahiti. We had a few hours together. That was our life: snatch what time we could to be together.

Bora Bora was a dream. We had a hut out over the water all to ourselves, with a wild lizard in it that we named "Eddie" for Eddie Izzard, one of my favorite comedians. Within the hut's floor was a small glass window to see the water below. Amy snapped a photo of me floating under the hut.

We took full advantage of the island: going to our own private beach, taking a cruise to feed the reef sharks and stand in the water with large rays, and visiting a black-pearl farm where Amy chose a pearl, my gift to her. Amazingly, I didn't obsess about the upcoming deployment. Amy made that easy.

In 2007, President Bush sent more troops to Iraq to establish a "unified, democratic federal Iraq that can govern itself, defend itself, and sustain itself, and is an ally in the War on Terror." Our role was to help Iraqis clear and secure neighborhoods to ensure they were capable of providing their own security.

By late June, I was deployed with 1/1 to Al Anbar Province, Iraq, my seventh deployment for clearing operations. I was the company commander of a group of 180 Marines, with my own executive officer and four platoon commanders. Knowing it was dangerous, Amy took my being there hard. Her worry proved justified.

In August, an IED placed and concealed within a road drainage pipe exploded, hitting a convoy of our vehicles from 1st Platoon. One Humvee sank down into the water. The driver suffered triple traumatic

VARPAS DE SA PEREIRA, PsyD

amputations and was unlikely to survive. The turret gunner, the corpsman, and the other Marine in the vehicle did what they could to render him and each other aid.

As soon as the call about the strike came over the radio at the company command post, I launched my quick reaction force (QRF), another platoon of Marines. They would provide security, establish an LZ for the medevac, and assist as needed.

They arrived in the area around midnight. It was too dark for the QRF to find the ambushed platoon who had reported enemy contact.

Overshooting a turn, the QRF encountered another IED. The squad leader and platoon sergeant were in the hit vehicle. Neither responded to radio calls.

I took my small personal security force (PSF), consisting of thirteen Marines, to assist. When we got to the QRF, we didn't find any enemy. We loaded up the casualties and moved back to the command post, leaving the damaged Humvee where it was. One of my corporals went to the back of the vehicle to "zero" all the systems in case the enemy got to the vehicle before we returned. The remainder of the QRF and 1st Platoon evacuated the site of the first IED strike.

A sniper team in the area reported having seen where the culprit for detonating the IED fled. I loaded my PSF back up, grabbed a HET Marine and his interpreter, and left to find the bastard who'd done this.

Finding the specific building that the snipers had identified took time. Finally, we entered and found an unarmed military-aged male there. My HET Marine, using his Arabic interpreter, drilled him to find out the person responsible for detonating the IEDs. He said little. To get to the bottom of it, we brought him back to the detention area at our CP.

By this time, it was dawn. Needing to recover the damaged vehicles, we sent a motor transport recovery unit to pull the wrecks out. Getting the first vehicle out took most of the day. Then, we escorted the recovery team to the second wreck. After loading it up, we headed back to the CP with the QRF and the recovery team in a column. In a position a few vehicles behind the QRF platoon commander and not far from the house that we'd breached, we hit another IED.

The turret gunner was killed instantly, the driver disoriented, and the vehicle commander badly burned. I ordered my PSF to secure the area.

I was riding in one of the few mine-resistant ambush-protected (MRAP) vehicles in the area. Taking it out would be a prize for the enemy. While my corpsman tended to the burned Marine, the enemy took their shot. I heard the boom while talking to some of the recovery team. I thought a large electrical pole transformer must have exploded from an overload, but the sound was behind me. That seemed odd. Within a second, I heard another RPG. "RPG!" shouted one of my Marines.

I took cover by the wheel of my MRAP. Following the RPG shots were small arms (rifle) fire from the north side of the canal. That area did not belong to my company. It was supposed to have been cleared by the U.S. Army 5th Battalion, 7th Cavalry, but no one was there to help.

Battalion called me. They were authorizing a "crossbow" (cross-boundary), meaning we could go into the Army cavalry unit's zone and take care of those ambushers. Though we didn't know whether the small bridges would support us on foot because of how heavy we'd be while wearing our body armor, we had to chance it. Our immediate concern was suppressing the enemy fire.

I kept as low as I could, moving from Humvee to Humvee so as not to get my head shot off. Images of Amy at my funeral popped into my head. Quickly, I stuffed thoughts of her out of my mind and directed fires. The enemy was close by, in reeds on the other side of the canal. Our machine guns on top of the vehicles couldn't drop their barrels low enough to shoot at them.

"Take out the rear pin and go free gun," I told the gunner.

"Still can't get it low enough, sir!"

"Use your rifle; get the pyro out. Fire star clusters into those reeds to smoke and burn them out to where you can engage with the machine guns."

I didn't have a forward air controller (FAC) or forward observer with me. Along with one of my sergeants, I would have to coordinate indirect fire assets that we could bear. I called a "troops in contact" into the battalion, meaning we were priority for everything.

The regiment offered to establish an artillery firing solution for us. I'd done my homework early on this deployment and knew that we were at the maximum effective range of those cannons at our location and dangerously close to anything they'd shoot. So, I wasn't willing to take the risk.

"We've got a section of Cobras on standby," they said, referring to helicopter gunships. I took those. Though I'd have to relay terminal guidance through my FAC at the CP, that would be easy. The AH-1 Cobras, attack helicopters, could provide us with long enough fire support and wouldn't be hindered by the canal. I cleared them "hot" to engage anything north of the canal. Providing us with what support they could offer from up in the air, they destroyed a vehicle full of combatants fleeing the area.

We weren't yet out of the woods. The enemy was still hiding in the reeds. We ended up unable to "close with and destroy" them "by fire and maneuver," as the Marine Corps purpose of the rifle squad dictates.

The tally for that day and a half was costly. Two killed in action, one very seriously injured, four seriously injured, seven moderately injured, and three lost HMMWVs, or Humvees. It wouldn't be the last time that we were shot at during the deployment, but the action would never again reach that level. We had our memorials and then went back to work.

Luckily, I had a stellar commander for that op. By the time I'd gotten back to my command post, he was there with members of his staff to man the radios and positions and do whatever was necessary so that my men and I could take a break.

"How long have you been up?" he asked me.

"Well over seventy-two hours by now, sir."

"Go to sleep. We'll handle everything for a while."

I did, grateful for his order to rest. With bad guys still out there, I would have continued until we got everyone. Despite having been hit with artillery fire and mortar fire and shot at with RPGs and AK-47s, I still hadn't *seen* a bad guy. My thirst for war was still not satisfied.

On this occasion, I hadn't lost any close buddies. Those who did, I knew, would feel worse and hunger even more for the enemy's blood.

I had to prepare for the memorial services of these Marines, which we would have in Iraq, and I needed to make sense of it. I combed through any material I had about losing troops in combat. Little was written about it at that time. I had one article written by an instructor at MCCC. He'd lost Marines in Fallujah with 3/1 and had jotted down his thoughts for others to consider. They were helpful. Later, I sent him an email to let him know that I appreciated what he'd done.

Not until the combat ended did I think about Amy. I knew she would find out from the other wives what had happened. Hiding this from her would be impossible.

My first sergeant got through to his wife. She had spoken to Amy, who had told her to tell me she was "solid"—a code for us meaning she didn't need me to call right away. I could focus on the next actions. My dear wife, I later found out, took it upon herself to check on my wounded Marines at the naval hospital in Balboa, an hour south from where we lived. Would she also feel compelled to attend all of the funerals? She did, flying on her own dime.

During the course of this deployment, I was a finalist for the award for the top infantry captain in the Marine Corps. The winner was a captain who was stuck in Okinawa. To console me, Amy told me how thrilled she was that I came close to winning.

Later, I found out I was passed up because they needed to "share the love," as too many commanders from combat units were getting the award. Still, the next MEU commander knew me by name and was pleased that I would be one of his maneuver element leaders. After feeling insulted, demeaned, ignored, and unappreciated in my first years of deployments, I finally felt vindicated—even more so after an article was published that included my role in October 2007 to restore peace in Iraq.

At that time, I was commander of the Marines of Company C, 1st Battalion, 1st Marine Regiment, Regimental Combat Team 6, in the city of Saqlawiyah. That experience was published online through the Marine Corps Public Affairs Office. The article, written by Corporal Bryce Muhlenberg, was titled "Ready to Fight Battalion Enables Transition in Anbar: Regimental Combat Team 6 Released on September 7th, 2007."

In the article, Muhlenberg quoted me: "In three months, we went from a place where there was pre-dug IEDs, sniper fire, weak security and the Army was taking fire constantly . . . [to] a place where the people are back in the marketplaces, back to their jobs and lives, and back to living a safer life altogether."

The article then stated that this transformation was a welcome relief to the Marines and Iraqis who had spent the last three years working together to make life in Iraq safe and streamlined. Citing me, Muhlenberg went on to write that this seed of success had been planted using an important strategy based on three simple principles: clear, hold, and build. It had been used by the Marines of Company C every single day since their arrival.

He then quoted me again: "We have worked off of the success of Operations Street Sweeper, Street Sweeper II, and Texas, which cleared the area of insurgents and the dissension they bring to the people of Iraq."

With the clear and hold phases complete, 1st Platoon was now ready to start the building process. Again, he quoted me directly:

> Soon we will transition completely, hand the security over and just provide overwatch. We won't man checkpoints, they will handle the security completely and we will work more on governance. We will be relying heavily on the Iraqi government to take responsibility for the building process.

The article then went on to state that in an area once considered Iraq's "Wild West," there was now peace, progress, and provincial security.

Amy emailed a copy of the article to everyone we knew.

While it gave me satisfaction to transform this war-torn area into one where people could resume a more normal life, I hungered for more. It was during the first month or so of deployment that the Marines in my command had been killed or wounded. Now, danger and risk were gone. The remaining time would involve more politics than tactics. I wondered how I could get back to combat.

Then, an incredible opportunity opened up for me with JSOC.

This had been my dream. Being a troop commander in JSOC was the ultimate machismo high. Masters of espionage trained to take on hijackers, terrorists, and enemy armies, special operations units can deploy by parachute, survive alone in hostile cities, speak foreign languages fluently, and strike at enemy targets with stunning swiftness and extraordinary teamwork.

Movies have been made about special operations units. In *Killer Elite,* Jason Statham's top-secret U.S. Army special operations unit runs covert missions all over the world, including the capture of Saddam Hussein and the location of al-Qaeda leader Osama bin Laden.

The JSOC units were deploying at a rapid rate, including for more direct action strikes against targets. I had to be a part of it. But JSOC was Army and Navy, and I was a Marine. Could I join?

As it turned out, I was branch-qualified because I had been a company commander—I could be a commando after all. Finally, my career would take off in the direction I had worked so hard for. Now that I had completed the Iraq deployment as a company commander, I could go through JSOC's rigorous assessment and selection program.

How exciting was *that*!

I thought that one tour as a company commander was enough for me to be branch-qualified. I contacted the Marine liaisons so I could apply to JSOC.

I got a lot of pushback: "They don't want you," "You're not going to make it," and so on. Some of this was professional jealousy. Marines, SEALs and Army Rangers are jealous of JSOC operators because of their operational tempo and mission set. Some of the pushback was cultural: the Marine Corps frowns on "special operations" because the presumption should be that every Marine is just as capable as the next. Other challenges to my ambition involved bureaucracy and procedure. While soldiers selected for JSOC would be moved right away, this would be an inconvenience to the Marines, so you could only try out when you were normally due for orders, or about once every three years.

"They don't know you," Amy consoled me. "They don't know what you're capable of."

To try out for JSOC, I had to be within a year of change-of-duty orders. That meant I would end up completing another deployment with 1/1, beginning in 2009, this time as the weapons company commander and fire support leader. We were part of the 13th MEU headed for the Persian Gulf. As the leader of the fire support center (FiSC), I was not likely to shoot my rifle. All I could do was dream we would be sent into harm's way. As it turned out, I came close.

Our ship, the USS *New Orleans*, collided with a submarine, the USS *Hartford*. While our ship was being repaired, we spent our time ashore in Bahrain and Djibouti. The rest of the MEU joined the counter-piracy operations off the coast of Somalia and participated in the rescue of Captain Phillips from the *Maersk Alabama*. The MEU was set to assist with the recapture of a German-flagged vessel, as well. However, at the last minute, the German government abandoned their plans.

My son, Orion, was born on April 11, 2009, while I was in Bahrain. Luckily, I had access to a phone. A nurse put my wife's cellphone on speaker mode, and I followed the labor and delivery, consoling and supporting my wife as much as I could from the other side of the world. Later, I learned through an email that he might be deaf and had to be fitted for hearing aids when he was three weeks old. I felt I had betrayed

Amy by not being there. I wouldn't see Amy and our baby until later that year, when they flew to meet me after our MEU returned home.

While we were sailing back to San Diego from the Middle East, we stopped at Pearl Harbor for a weekend. My son was four months old, and Amy flew out with him to Hawaii. It would be the first time I met him. After that weekend, the MEU would sail for a week from Oahu to San Diego. I asked for a leave to stay in Hawaii with them and then fly home.

I was told, "I would have thought you would want to finish the deployment with your men." I didn't insist. The hierarchy was intimidating. They would joke that if the service wanted you to have a family, they'd have issued you one. "Family second" was the model.

I didn't see my little butterball baby son again until I returned with 1/1 to Camp Pendleton in California in August. An eating, crying, defecating biological thing, Orion didn't seem real to me, and I didn't immediately bond. Not until he was a little older and developed a personality did I see him as more my son rather than an anonymous, tiny, chubby human.

In September, after only a month with my family, I reported to 1st Marine Regiment at Camp Pendleton as a staff officer. I had to leave the responsibility of home and a special-needs infant entirely in Amy's hands, although she herself had a full-time job. Luckily, her mother was available to help out and babysit during the day.

When I got back from deployment in August 2009, I was promoted to major and got time to work on getting to JSOC. Typically, Marines receive permanent change of station or permanent change of assignment orders every twenty-four to thirty-six months in order to open and fill billets as others progress in their careers. Because I'd been in the 1st Marine Division for three years, I had to take a new set of orders. Luckily, I was moved locally to the school of infantry (SOI) at Camp Pendleton with my wife and child, with the understanding that I'd take a month to screen for JSOC in 2010.

With the collapse of discussions between the U.S. and Iraq governments about extending the stay of U.S. troops in Iraq, in October 2011, President Obama announced the full withdrawal of troops as previously scheduled.

Amy was in a predicament. Having a kid and a full-time job was hard on her. If I were chosen for JSOC, we would have to separate for a while. She would have preferred for me to retire or try to get local orders for California again. On the other hand, she had encouraged me to try out for JSOC. If I didn't go through assessment and selection, I might hold "shoulda, woulda, coulda" against her.

The solution would be for Amy to move to where I was stationed. We considered this, but unfortunately, the positions available to her as an engineer on the East Coast in the tech industry were few and far between. Ultimately, we decided she would stay at her job in California, with her parents there to help raise our son, and we would revisit our situation after my year of training was done. If I didn't pass the operator course, then there would be no harm done. I would just get orders back to the West Coast.

In the fall of 2010, I passed the selection for JSOC. Amy was pregnant, and I owed the SOI the remainder of 2011 to teach the up-and-coming combat instructors before I would be allowed to move. My daughter Alessandra was born in July 2011. I was authorized ten days of time off for the delivery and bonding time. Having missed part of Amy's pregnancy at selection and time in the field at SOI, I was thrilled to be there to drive Amy to the hospital.

I was finally with my family. I went to work early for physical fitness training and was usually home for dinner. Spending time with Amy, Alessandra, and Orion, I felt more like a family man and a little less like a warrior. It didn't feel great, to be honest. Where was my chance to kill bad guys? I wouldn't be able to start with JSOC until January 2012.

The Marine Corps made me move to the East Coast early in November 2011, when my time was up with SOI. My orders to JSOC would only be for three years, and being done with SOI started my clock with them.

The operator training course wouldn't begin until January 2012. This meant that I would lose two of the thirty-six months I'd spend with JSOC milling around and doing nothing. I asked the administrative officer, nicknamed "Smitty," if he could modify my orders so that I could start at JSOC in January.

"Sorry," said Smitty, the quintessential bureaucrat. "There's nothing I can do."

I ended up flying to the unit, checking in, and then flying back to California on leave for two months, returning in January to start the operator training course.

The majority of my training was conducted on the East Coast to be in the same time zone as Washington, D.C.—where, specifically, no one can know. Because of the special nature of JSOC, the unit locations and where the courses operate were classified to protect JSOC operators.

Depending on the JSOC unit, training was tailored to their specific mission set. All units have weeks of counterterrorism driving, special-purpose austere medical courses, fieldcraft, some tradecraft, full-mission-profile training evolutions, and thousands and thousands of fired rounds.

In December 2011, I said goodbye to Amy, two-and-a-half-year-old Orion, and six-month-old Alessandra, and I returned to JSOC to spend the next year in the operator course. I would have only seventeen nonconsecutive days off until Christmas.

Amy and the kids visited as often as they could, though that was hard for my wife. It was a challenge to fly back and forth while managing two little kids on the plane and in the airport. These trips also meant that she had to cut back on traveling for her work, a step that might interfere with her earning a promotion. Still, we had no other choice. It was impossible for me to fly west and be back in time for training.

At the end of the operator course, Amy told me she was pregnant again and cursed my "potent swimmers." Being pregnant with two little kids while hemorrhaging money trying to maintain two households in California and the East Coast was more than she could manage. She pleaded with me not to take the next step to go to the operational squadron when I graduated.

"What?! I've trained for this my whole life. I can't now stop and not operate!"

In a no-win situation again, Amy knew I would resent her if she was the reason I stopped. Still, she held back her approval until she could check out my squadron commander, knowing what grief I had been through with Marone and Clay. She got the opportunity at a family event when I checked into the JSOC squadron.

I introduced her to Lieutenant Colonel O'Leary. Not a man in need of flexing his muscles, he kept his hair short and hid his athleticism under larger-fitting clothing. What a contrast to those assholes Marone and Clay!

O'Leary greeted Amy with a big, toothy smile and put his arm around my shoulders. "You've got a sharp, ambitious husband here," he said. "We're happy to have him."

"Thank you," she replied. Her face took on a serious tone and she said, "I've got some questions."

His smile faded a bit as this was supposed to be a casual event in a bowling alley.

They sat down, and she peppered him with questions. How would information get to me when I was "down range," or deployed? Who would be the arbiter if she stated she had an emergency in California and needed me home? How long would it take for them to get me back? How much would she know about where I was and what I was doing? Her questions went on and on.

Unbeknownst to her, her queries hit the nail on the head. One of the soldiers from my assessment and selection class had died down range. His role in that foreign country was unclear to his family, and not getting the full story, they had thought he had joined some rear echelon unit and wasn't in danger.

He had died under O'Leary's watch. Now Amy hurled questions at him regarding just such an event—that I, too, could be in a dangerous place, and she wouldn't know.

To avoid swaying the answers, I walked away so O'Leary couldn't look to me to bail him out. Amy is a great judge of character. Letting her corner him one-on-one worked best. In the end, he allayed most of her fears and got her approval.

She had less luck in connecting with the wives. Amy is a professional. Her work defines her, not her marriage. Apparently, the opposite was true of some of the wives. They sacrificed all for their husbands, quitting their jobs and moving without complaints to the East Coast to support them.

Looking askance at Amy, one wife said, "You just have to leave your job."

"I don't have that kind of a job; I'm a professional engineer at a consumer electronic company."

"Well, I was a nurse. You just have to give it up to support your husband."

The woman glanced at the other wives, who nodded. "We all did."

Amy didn't share that she actually hoped for the opposite arrangement. She wanted me to give up JSOC after my year in the operator course in 2012 and come back to California.

"It's too much for me to handle the kids, the house, my job," she would say. "And we don't have enough money to cover our expenses."

We had two houses going, one in Orange County and one on the East Coast. Now, with the addition of our third child, Celina, born on June 29, 2013, we were going broke, something Amy hid from me until my last year with the unit. Even with her large salary and mine, so much of our money was being spent on childcare, rent, and a mortgage that we were living paycheck to paycheck.

I thought that the specialist pay from JSOC would cover my cost-of-living expenses. It did not. I was paid a few hundred dollars extra a month for language capability, demolitions work, and airborne operations—not enough to pay the rent where I was living.

On top of this, Amy's parents had stopped assisting us with childcare, and Amy had to search for a nanny in the community. Finding someone who would watch three small children wasn't cheap. Additionally, Amy still had to travel for work.

"You don't understand," I said to her, trying to explain how stopping right before I attained my goal would be devastating for me. "This is like I've been training for the Superbowl my whole life, and now I'm not going to be able to play!"

What could she do but agree to let me continue?

Another reason we were running low on money was because of my air travel. Every time I'd combine a flight to the Far East or to the special operations command in Hawaii with a visit to Amy and the kids, the government was not paying for my full flight and, at times, shorting me between two to five thousand dollars per trip.

One of the sergeants major reminded the squadron commander that I had to take leave to see my wife and kids. Apparently, it had come up during a meeting about whether or not I should be sent overseas more or be selected for a training mission. Out of the training course, I was put right to work. And several times, I was geared up to go down range as part of a task force for four months. While this assignment delighted me, I was still trying to deploy.

CHAPTER 9

ON THE ROAD TO RETIREMENT

Old soldiers never die; they just fade away.

—General Douglas MacArthur

To push my career forward, I wanted to be promoted to lieutenant colonel, just below the rank of colonel. This required completing the Marine command and staff college. I would have to complete this academic work in addition to the operational work I was doing with JSOC.

In 2013, I started a blended program, including some online and some resident learning. While I could find time to do the resident portion at the Marine base in Quantico, I had to take permissive leave from JSOC, because I had to report to Quantico every day for a few weeks. I was frustrated, because taking the course prevented me from going to Yemen and then deploying to Afghanistan.

Yemen was a melting pot of different factions, from al-Qaeda elements to the Houthi rebels. JSOC was working with the Yemeni government to manage national security concerns. Yemen provided JSOC with a place to launch operations into other nearby countries, as well.

My role there would have been to work with the Yemeni government in a counterterrorism capacity, directing their SOF to try and kill or capture key leaders. My other responsibilities would be to monitor and interdict the piracy around Somalia and Yemen. The situation in Yemen continued to deteriorate until, in 2014, the Houthi rebels seized control of the capital and much of the North. American boots on the ground were no longer authorized.

For most of 2014, I worked on taking control of the task force (TF) unit in Afghanistan, a position that, as a major with an O-4 pay grade, I was suited for. Command of a TF is usually reserved for a lieutenant colonel (an O-5). I was nominated to lead the task force by some of the senior staff at special operations command central. Combat there continued even to the end of U.S. involvement in Afghanistan, when TF forces were still finding, killing, or capturing Taliban and al-Qaeda leaders.

While I was preparing to deploy to Afghanistan, the Marine Corps determined that I should attend command and staff school as a resident. This meant that HQMC wanted me to repeat the same coursework I had completed in the blended program and that I would retake the course as a resident. The resident course lasts one year, and they wanted me to start after I'd only been with JSOC for twenty-four months. My Marine administrative command—good ol' Smitty—could, of course, do nothing.

"Well, sir," he said, "it looks like your only options are to resign or take the orders to the school." With the first option, he meant that if I refused orders, I would essentially have to resign, or leave the military.

"Those can't be the only options. What if I leave the Marines and switch to the Army?"

"I guess that might work, sir."

"What paperwork do I need to fill out to cross-deck [leave one branch for another]?"

"I'm not sure, sir. I can look into it."

Meanwhile, I wouldn't take his word for it. Someone had added my name to the list of those who were going to school. To find out who, I trekked to HQMC in Quantico. I arrived in street clothes, in need of a haircut, thinking my status as a JSOC operator rather than a Marine might work to my advantage. Because operators work largely incognito, their look has relaxed standards. Some of my Marine counterparts at JSOC would shave their hair into mohawks and grow beards, something regular and Recon Marines are not allowed to do.

"How does this work?" I asked the guy who released the list. "Because I see that three guys that were in my same class year aren't on this list."

I listed their names. One was an F/A-18 pilot. The other academic officer recognized the name and came over to join the discussion.

"You know Perp?" asked the officer. "Perp" was the pilot's call sign.

"Yes, we deployed together."

"Yeah, when we got the initial list of who was eligible, I saw that his name was on there, and I crossed it off. He's with MARSOC right now, and if he did another year in school, he'd lose his jet qualifications."

"I'm in MARSOC right now. Can you cross my name off, too? I'm doing the blended program anyway and will literally walk in the same graduation ceremony as these resident guys."

"I'm sure we can figure something out."

I went back to JSOC having scrubbed my name from the resident list. I would be free to deploy again.

In March 2014, I was sent to Germany to conduct a brief to the supreme allied commander of European forces. I was in charge of an operation, within his purview, to track a terrorist. I needed his signature to proceed with the classified mission against that person and his terrorist organization. This was a big deal. I practiced the brief several times with my commander and his commander, who both flew with me to Germany in case I fucked it up.

Right after delivering the brief successfully and adding feathers to my cap, my spirit was then crushed: no sooner did I get back to my hotel than I discovered I was assigned orders to Camp Lejeune, North Carolina, and not Camp Pendleton in California, where my family was. Knowing I was going to report somewhere away from them at the end of 2014, given that my time in JSOC would be up, I had submitted my choice as Camp Pendleton. Amy and the kids lived just north of the base, in the same house we'd lived in while I was stationed at Camp Pendleton from 2008 to 2011.

I was devastated. I had no idea what to do, and neither did my Army commanders at the unit. They deferred to the Marine liaison and the special operations program directorate (PO-SOD) at the Pentagon.

I called the Marine who would be my commander in North Carolina. I knew him from our time at FAST and 1/5. He didn't seem thrilled with my assignment to him, either.

"I was hoping for someone more conventional," he said.

Hmmm . . . my reputation had gotten around!

I asked the monitor why he cut me to a geographic location not even on my list.

"That's where the vacancies are," he replied.

"There are three other guys you're sending to Camp Pendleton. I have seniority over all of them. Why didn't I get to choose ahead of them?"

"Needs of the Marine Corps," he said. Ostensibly, this bullshit explanation meant that I'd spent most of my career with 1st Marine Division in California, and to "round me out," I now needed to spend time in North Carolina with 2 MARDIV. In reality, it meant the monitor was a chickenshit and didn't want to tell me the real reason I didn't get top choice.

"What about openings for staff officers at Camp Pendleton?" I asked. "I'm willing to take any job there to get back to my family."

"You need to go back to an infantry battalion as an OpsO [operations officer] or XO, or you won't be competitive for promotion," the monitor told me.

"I'm already on the board for promotion to lieutenant colonel; I'll either make lieutenant colonel or I won't. My years as a troop commander in JSOC should guarantee that promotion."

"You still need to have one of those billets [jobs] or you won't be competitive for battalion command."

"I don't want to take battalion command."

"Well, it's my job to look out for your career and make sure that you're competitive."

"So, by looking out for my career, you're effectively ending it?"

"That's your choice."

I wanted to punch that fucker in the face.

I checked with other Marines in the special operations directorate (SOD) about alternative options for me. Their solution was to trade the North Carolina orders for ones at the Marine Corps Air Ground Combat Center in Twentynine Palms, a base located in the California desert and 150 miles, or a three-and-a-half-hour drive, from where Amy and the kids were living. The colonel at SOD thought that at least that was 1st Marine Division (1st MARDIV). One of the infantry regiments of the division, 7th Marine Regiment, is based at Twentynine Palms in addition to a few other Marines units, while the majority and headquarters of 1st MARDIV are at Camp Pendleton. He posited that as a major, I'd check into the division headquarters at Camp Pendleton first, and then I could plead my case and be able to change the orders within the division. As more than three-quarters of 1st MARDIV is based at Camp Pendleton, this sounded like a reasonable plan.

The next colonel in charge of SOD told me they could not modify those orders and that the previous colonel was mistaken. I would need to check into Twentynine Palms, not Camp Pendleton.

By this point, I didn't have the finances to travel back and forth between my family and Twentynine Palms for another three years. I considered the Exceptional Family Member Program (EFMP). Both of my older kids, and now all four of my kids, had some amount of hearing loss, from profound to moderate. Orion was awaiting a cochlear implant. By getting into the EFMP, I would be able to shift my orders to Camp Pendleton. I argued that I was eligible because there weren't enough providers out in the middle of the desert for my kids.

HQMC denied my claims because without considering the quality of care, they found some random provider within fifty miles who had availability and took TRICARE health insurance.

Not accommodating me made no sense. I was one of five officers who could do some of the more challenging jobs of managing operators under special access programs. Especially after the Russians invaded Crimea, there was concern about who could run operations in the Baltics. At a conference in Stuttgart in 2014, I said, "I know the guy. It would be me. But you need to authorize those operations now, because it'll take me two years to build what you're looking for."

Even the SEALs I worked with didn't understand what was happening to me. One SEAL commander told me that EFMP was a good program; it let him stay in San Diego for his entire twenty-year career!

I couldn't figure it out. Here I was, an operator at the tip of the spear, basically begging for a job at Camp Pendleton where I could keep my marriage and family together, and I was being denied.

I spoke to the colonel at SOD and my administrative command in Quantico. They said there was nothing that they could do.

"So, my options are Twentynine Palms as a geographic bachelor for another three years, or refuse the orders?"

"Looks like it, sir," said Smitty.

"Fine, I'm putting in a package for early retirement."

I submitted the paperwork for my retirement. Because President Obama was drawing down the military, I could retire at 15.5 years and immediately draw a pension.

When I submitted my paperwork in the summer of 2014, I thought I would finish my time with JSOC, bringing me to the end of 2014, and then take up to March 2015 to set up my classified special access program (SAP).

This program would allow me to work with JSOC as a civilian. The SAPs carry the highest security clearance. With my skill set developed through my time as a commander at JSOC, I could work as an

Army civilian in one of the programs I'd been in charge of for the last three years.

I could effectively trade my uniform for civilian clothing. I'd been mostly wearing civilian clothing as a JSOC guy already. My "home base" would be in California, and I would deploy with the SAP as required. Similar to contracting for the CIA or private security firms like Blackwater, all that mattered was that I lived close to an airport and could get to "the target" or area of operation within a few days.

To make all of that happen, I asked for my retirement to be set for March 2015. Typically, whatever date you ask for, you get. Since I was submitting my retirement around September 2014, I figured that six months to prepare would be reasonable.

I got back my retirement date as December 2014. The manpower branch at HQMC had deliberately pulled in my retirement date. In the minds of those personnel administrators at HQMC, I was trying to delay retirement in order to see the results of the promotion board for lieutenant colonel. If I was selected for promotion, I could pull my retirement papers and demand a new set of orders. In order to prevent that possibility, the retirement date was set prior to the results being posted.

I had no intention of doing that. I was done with the Marines, and the results of the board didn't matter to me. I needed the time to complete my medical out-processing while trying to figure out how to get my kids on my GI Bill for their college use and other administrative concerns. I also needed that time to ensure I was brought onto the SAP.

Because my JSOC unit was run by the Army, I had a Marine administrative commander in charge of me "on paper." I was loaned out to JSOC, where day to day I worked for an Army colonel.

Smitty worked for the Marine administrative colonel. As my retirement in December 2014 approached, Smitty pushed me to sign papers that would make his life easier. If I signed on the dotted line, he could push the paperwork without any fuss. Neither my Marine administrative commander nor Smitty wanted me to fight the retirement date.

They didn't want me to "request mast," which meant speaking directly to the commanding general of MARSOC about my situation. All of that meant more work for them. It hit me then how ugly a departure from the military could get, which, I would find later, was a common experience among the veterans I treated for WWS.

To avoid more paperwork, the colonel, who was supposed to be my advocate, built a case against me. At the same time, the colonel was so delinquent in paperwork that my GI Bill requests lapsed—this was the last straw. I demanded to speak with the two-star general in charge of MARSOC.

I wanted to see him in person, so I trekked to Camp Lejeune, where he was stationed. Once there, I found out I could only have a video conference with him. I had no choice, though I knew it was my right to see him face to face.

While waiting for that to be set up, I called into HQMC to fix the GI Bill paperwork. I finally found someone helpful who, unlike the administrative commander, understood that what mattered to me was when exactly I signed the forms. And so my kids were added to my GI Bill.

Finally, I got to speak directly to the commanding general. I showed up to the video conference room in my Marine uniform, which I hadn't worn for a long time. Alone with the commanding general on a video screen, I explained to him the challenges presented by my family's medical issues. He merely affirmed that he would not change my retirement date. He pointed to the success of getting my GI Bill fixed as evidence of how much hard work they were doing on my behalf. He wrapped up by throwing out some platitudes about my service, believing everything had been taken care of.

It wasn't. I had hit a wall. Although the JSOC commanding general outranked the administrative commander, he didn't want to get involved in "Marine matters."

The next day, the administrative officer brought me a form to fill out and sign, which stated that I felt everything from the request mast had been resolved.

"I'm not signing this. Nothing was resolved."

"Sir, you just have to sign this form so we can close out the request mast."

"The matter isn't resolved, and I don't know what steps the general has taken, if any, on my behalf. So no, I can't sign that, as I'm unsatisfied with the process, nor can I acknowledge that I understand what the outcome and resolutions were."

"Just sign the form, and then you can call and talk through those concerns."

"No, I won't."

Things continued to go further sideways. The administrators at HQMC audited my records and found that they'd overpaid me every month for fifteen years and that I owed the government about $60,000. So they stopped sending me paychecks. Eventually, I was found not at fault for those overpayments by some finance guy at HQMC who determined I wouldn't have to pay it back. While grateful for the debt being erased, I found it incredulous that this was how I was being treated in my last month as a Marine.

I was sick of it. I didn't know what I was going to do next. I had offers to go work for the CIA, but that meant more time away from Amy and the kids. Near the end, I pushed for more work with JSOC in a SAP. That didn't work out, however, because JSOC was in the process of terminating programs like the one I was involved in. I was offered a different SAP to work on that would require constant deployments, more frequent than even those of active-duty personnel. The constant time away would be too long for me to keep my family going. Furthermore, this kind of down-range deployment would prevent me from calling or contacting home.

I offered to participate if I could shorten the deployment lengths by about 33 percent and make up for it by doing them more frequently. While the manager and I knew that this would make sense and reduce the burnout of the operatives overall, the program itself was on thin ice after security failures involving similar programs.

By January 2015, I was adrift. Amy was traveling with her job, and I was Mr. Mom full-time. In the fall of 2015, Orion started kindergarten, and I volunteered in the classroom as a parent helper. In October, I took him to a birthday party where the mom had arranged for a sheriff to show up.

After that, I thought about Orion's upcoming birthday. I knew he wanted a Marine birthday party. I set up a one-rope bridge the kids could use as a zip line, along with camouflage shirts, paint, dog tags, and so on. While the party distracted me from my underlying combat envy, it obviously wasn't enough to satisfy my need for purpose in life. I was obsessed with redeploying.

In 2016, I contracted part-time with JSOC as a civilian. I would help complete some assessments of new JSOC candidates. Most of these assessments required spending one to two weeks with the candidates and observing them in mission scenarios. This placed me around active-duty guys and gave me a high that I wasn't getting at home with my family. After work on a contract or an evening out in San Diego with one of the guys, I would get grumpy back at home, and Amy found me hard to be around. She says I'm getting better now. The altered, difficult moods now only last about a week, whereas before they would last months.

At that time, however, I was clearly in the throes of an identity crisis, common in vets returning to civilian life. Let's now explore this problem.

PART III

IDENTITY CRISIS

CHAPTER 10

HANGING UP THE UNIFORM

Treat the civilian world as a foreign country. They speak a different language, have a different culture and different customs. Adapt to the new country by adapting to the language, culture, and customs.

—Militarytransition.org

I was a wild kid, easily bored and always exploring. On any given day, you could find me in a tree or on the roof. To help me focus and keep out of trouble, I was drawn to discipline and rules, an important characteristic for my future military life. At five, I joined the Cub Scouts. I loved the order of the Scouts, and I liked wearing a uniform. The Scouts—and later, boarding school, ROTC, and the military—gave me the external structure I needed to stay centered.

Following rules was also natural for me because I grew up Catholic in a tight-knit Lithuanian community with strict rules and a tribal bond. This background made me an ideal candidate for the military.

Whether you are Army, Navy, Air Force, Space Force, or Marines, your identity is entrenched in following stringent rules and keeping strict discipline to create order, obedience, and commitment to your team. Regardless of rank, you wake up between 4:00 and 6:00 a.m., put on your cap and cammies, do your job, get back to base, and hit the gym, the MWR (morale, welfare, and recreation center), or the chow hall, whatever your routine is. Sure, it could get boring and repetitive, but you knew how to do stuff and where things were. It was familiar.

When you take off the uniform, your world flips upside down. Suddenly you are no longer a soldier, a Marine, a Navy SEAL, a Green Beret, or an airman with daily life controlled and mapped out. Without that order and identity, a part of everyday life seems chaotic and confusing. It's hard to define *who you are*.

In his memoir *Jarhead*, Anthony Swofford describes how, after leaving the U.S. Marine Corps following the end of the war in the Gulf, he found it hard to adapt to civilian life due to extreme combat-related PTSD. It felt, said Swofford, "strange to be in a place without having someone telling me to throw my gear in a truck and go somewhere."

To make matters worse, many civilians don't appreciate your service. Veterans of World War II came home to ticker-tape parades and young girls throwing their arms around them, kissing them passionately. The wars in Vietnam, Iraq, and Afghanistan were thankless wars. Many citizens wondered what we were doing there. There were no more ticker-tape parades for victories. There was less admiration of the vets who had fought in Iraq or Afghanistan. Fewer Americans saw them as heroes and defenders of their country. Even those who were heralded as heroes were quickly forgotten.

Not only were returning vets not being honored, often they would feel irrelevant in the civilian job market. In particular, many who joined the military out of high school found it hard to find a viable civilian profession.

Even with a college degree in math, I, too, initially felt irrelevant. I could teach math or, as mentioned, join the FBI, CIA, or law enforcement, but none of these were paths I wanted to take. I couldn't figure out what I wanted to do. I needed something else that I couldn't get from being a cop.

I wondered if the VA had job openings that might appeal to me. I spoke to the vocation and rehabilitation counselor at the VA but was told that their program wasn't meant for me.

"You've already got a degree," he explained.

"What do you suppose I would do with a math degree?"

"You could teach math."

"I don't have a teaching credential."

"Well, I can't approve you for this program."

Man, I love the VA, I thought. *Designed to help veterans by denying them everything.*

"You can appeal this through a VSO," he added.

"What's a VSO?"

"Ah, you can ask the front desk. I can't direct you to a VSO."

I went to the front desk, where they told me that a VSO is a Veteran Service Organization, like the American Legion or the Veterans of Foreign Wars. To appeal the decision of the VA, I'd have to solicit help from an *outside* nonprofit organization—fifteen years of service, and this was my "welcome home, Mr. Marine."

As I walked out, I wondered about those warriors with few skills. What kind of reception would they get? I shuddered at the thought. Later, when I devised a theory for WWS, the difficulty of finding meaningful employment and, more importantly, *purpose* for the returning vet would ring loud as a primary motivation for the desire to return to combat.

Part of the problem lay with the military's lack of providing a gradual process of integration back into society. One day you're a warrior; the next, you're a civilian expected to be like everyone else. Except everyone else was not trained to kill, to be suspicious of their environment, to use anger as a survival mechanism, or to be willing to die for comrades in arms. This war mentality was hard to turn off willy-nilly; becoming like everyone else could be challenging.

One retired airman I worked with at the VA bemoaned the length of time the transition period lasted. When I was transitioning out, the course was about a week. If you had a job offer in hand, you would be excused from four days of the transitioning class. When he left, "they did all the training in about a day, maybe a few hours." No way was he was going to learn anything of value in that short a period of time.

With their lives lacking meaning, some found that returning to the battlefield was the only way to find purpose in life, and so many of them tried to redeploy. Among those for whom it was not a possibility because of injuries, some committed suicide as a way out. And indeed, suicide is significantly higher in ex-military men than in the general population.

Lacking a solid path forward, my choices on a daily basis seemed endless and went far beyond the larger question of a viable profession. It applied to everything: which shirt to wear, which TV station to watch, which restaurant to go to, and on and on.

Everything felt wrong. The military had been my compass. Without it, I had no direction. I didn't feel like my own person. I wanted to be fighting. Being responsible for a family interfered with this.

Stephen Sondheim's lyrics in "On the Steps of the Palace"—from the Broadway musical *Into the Woods*—capture some of this identity confusion. The song is about Cinderella figuring out if she should go to the ball when she knows that she doesn't belong in a palace. She sings the following:

> But then how can you know
> Who you are 'til you know
> What you want? Which I don't,
> So then which do you pick,
> Where you're safe, out of sight
> And yourself, but where everything's wrong?
> Or where everything's right
> But you know that you'll never belong?
> You'll be better off there
> Where there's nothing to choose
> So there's nothing to lose.

Not knowing "who you are" until "you know what you want" reso-nated with my lack of identity, my missing sense of purpose at home. Being "safe" and "out of sight" but "where everything's wrong" reminded me of my identity in the military, which was well-defined but abnor-mal to me. Despite this alienation, being at home, "where everything's right" in terms of comfort and love, felt like a foreign place I would "never belong." Consequently, being in a place "where there's nothing to choose"—in other words, the military—seemed like a more natural option. In tune with Sondheim's lyrics, I felt I would have been better off back in the desert, sweating bullets, than being in a foreign land where I didn't belong, where never-ending choices created only confusion.

The love of my family should have helped. It didn't. I felt estranged from them. Amy and I had spent little time together in the nine years we were married, and we barely knew each other. Going out together for the first few weeks felt like a series of first dates. We lived in differ-ent realities. Amy felt stressed taking care of a home and kids while also working. I felt stressed because I had been shot at and seen friends die or lose limbs, and my life had lost purpose.

Nor could I relate to my five-year-old son and my daughters, now three and one, whose childhoods I had mostly missed. I hardly knew them, and they hardly knew me.

Things did not get better my first year after retirement. No longer a warrior, I was now Mr. Mom, with Amy being the breadwinner. Need-ing to catch up to her peers and avoid a layoff in one of the tech-crunch consumer electronic company drawdowns, she spent more time at the office than at home, and she now traveled overseas at least once a quarter to compensate for two years of no foreign trips due to my service keep-ing me from home.

Being Daddy-Uber, I would drive Orion and Alessandra to school and come home to care for Celina, now turning two. I felt more like a nanny than a member of the family.

Other than the fun of being a twice-a-week soccer coach for my son, my only interest was gaming. Playing video games was a great way to fast-forward through time, the best ones having immediate and enduring appeal to my warrior instincts. I wasted hours each day with shooting games like *Call of Duty* or *Battlefield*. If I had a PC, I played *League of Legends* or more immersive games like *World of Warcraft*.

At times, I rushed Celina to take a nap so that I could play video games, or I'd let her watch a Baby Einstein video to get back to the Xbox to play *Call of Duty*. If I didn't have the responsibility of taking care of her, I would have gamed nonstop until needing to pick the kids up from school. Fortunately, these games progressed beyond my abilities and comfort levels, as many required downloading additional applications or joining calls that I was unwilling to participate in.

While it might have helped to share the myriad of discomforting feelings inside me with Amy, who was an understanding person and good listener, I found it hard to do so with her or anyone else. Only my military buddies could fully understand that hell I had been through. The problem was that most were still on active duty and in the fight, with little time for the retired guy back home.

To encourage communication skills with family and friends when coming back from deployments, the military offers "return and reunion" classes. Amy and I attended them. They discuss the difficulties of going from a world of order and structure and intense meaning to one of chaos and lack of purpose. They focus on the reasons why some men were traumatized by leaving that controlled world.

Amy listened attentively and tried hard not to rile me by asking too much about what deployment was like, or to get angry with me when I snapped at her for not understanding my experience—as the discussions recommended. She had the patience of a saint.

While I, too, found the classes somewhat helpful for enhancing my insight into my communication problems, they had limited impact. As with many returning vets, my poor relationship with my family went deep. I felt unappreciated and irrelevant. What role did I play in my family? I was unsure. Did they even need me? Having functioned for so long without me, they had a routine independent from me. I became the "extraneous responsible adult," not specifically needed—and easily replaceable.

These feelings had surfaced even before retirement. While in JSOC in 2013, I had two weeks off to be with my family. As soon as I got home, my wife asked me to build an outdoor shed as a playhouse for my four-year-old son and two-year-old daughter for a Christmas present.

"Sure." I figured it couldn't take too long.

I found a place where I could buy a little home constructed with two-by-fours and windows. They advertised it as easy to put together in a week or so. That was okay since I had two weeks of vacation. It sounded like all I would need was a screwdriver.

The "kit" arrived with some lumber. But before you could build it, you had to dig a large hole for a foundation. I spent all day digging the hole, using a pickax to get through all the roots, until it was six feet long, five feet wide, and one foot deep. Digging this foundation took almost all day.

Amy came out when I was finished doing this first step. "What's taking so long?" she asked.

Is she kidding? I thought. "Look," I said, "take one of the rose bushes we were going to plant and dig a small one-foot hole for it." I figured once she saw how hard it was to even dig a small hole, she would appreciate what I had been doing all day. She finally did.

I ended up needing a thousand pounds of gravel, pouring one hundred pounds at a time for the foundation. It soon became clear I would be unable to build the playhouse on my own, and I asked some military buddies to help. Two answered the call.

Still, we needed some skilled carpentry. I watched YouTube videos on how to put shingles on a roof. Ultimately, it took my two buddies, me, and two hired hands one week to finish the project by Christmas Day. My father-in-law finished the inside later by installing legitimate flooring. In the end, my son and daughter loved it.

But I was steaming. I didn't come home to do a "honey-do" list; I wanted to be with my kids, with whom I ended up spending very little time. This playhouse project ruined my vacation. In fact, I don't like doing domestic jobs that require much physical work. I have enough physical work in the military.

I felt unappreciated. Rather than a hero's welcome for the husband and daddy who was willing to sacrifice his life for his family, I was the handyman. In Iraq, I had to burn feces out of buckets. I did this literal shit work because it was my job. At home, I wanted to be pampered, not forced to do more shit work. I couldn't wait to return to base.

Feeling insignificant intensified when I finally retired. In the military, I had been in charge of 150 men who had to listen to me. At home, neither my wife nor my kids had to salute, say "yes, sir," and obey my orders. I felt diminished and emasculated.

What was the point of going on? Before, I went to the gym because it made a difference when having to carry a pack and heavy weapons. Strength and endurance made me a better combat Marine. What was the purpose of going to the gym now? Before, I would do marksmanship because the better I was at shooting, the better I'd perform down range. Why go to the range now? I wasn't gonna be shooting bad guys anymore. And the range costs money! Why lay out needed funds to maintain a skill I'm not going to use.

Why even get out of bed? I asked myself.

I had the "why bother" about everything, from shaving to putting on clean clothing. As it turned out, these were signs of clinical depression. Would I admit it at the time? Never.

Suspicious, tightly wound, and quick to lash out, I fought with Amy and snapped at the kids. I smashed my son's LEGO sets and threw objects the kids were fighting over, knowing I would destroy the toy, remote, or whatever it was. Even a petty squabble about who took the waffle from the toaster set me off. To stop their bickering, I swept the toaster off the island in the middle of the kitchen so forcefully that it flew through the air and dented the refrigerator.

"Dad!" said Alessandria, eyes wide with terror.

"*Shut up!*" I yelled.

She did.

I knew my autocratic behavior was harmful. The only thing this approach teaches is "might is right." Kids may behave, but they do it out of fear and learn nothing about how to modify their behavior in the future. So why did I go ballistic? In hindsight, my mind was yelling, *I quit for this? This is what my life is now? I get no respect from these smaller humans!*

Of course, "might is right" is the military cry. My identity was that of a warrior, not a wise father who disciplines his children through reason. And so, like a good warrior, I continued to yell at them, because that was effective in the short run. They feared me and obeyed. Why stop?

My outrageous behavior knew no end. One day, in front of the kids, I told Amy to "leave me the fuck alone." Her mouth dropped open. My son's eyes widened. Alessandra said, "Ohhhh . . . you said a bad word."

I felt mortified. If my mother had been there, she would have slapped me. Growing up, we never swore in my house. Only a small man needs a big word like that to convey their meaning, my mother used to say.

Not until I joined the Marines did I possess a foul mouth. Swearing and crudity are rites of passage in the Marines. It's how we relate to each other. The meme or joke about the Marine Corps is that we are ahead of the curve in terms of diversity: every race and creed of Marine is just another "motherfucker." Swearing is a mark of our natural anger.

In the little time Amy and I had spent together, she had not seen this ugly side of me. Nor had I, outside of my time in the Marines. Before Iraq, I rarely felt or expressed anger. I was an even-keeled guy, rarely overcome with emotions. Now, in order not to provoke my "Yosemite Sam" (or "Incredible Hulk") hair-trigger temper, my poor family had to tread on eggshells.

"He's just angry," she would tell the kids. "Give him space."

Meanwhile, the kids would assume it was their fault. I was failing as a Marine and as a father.

Even when I attempted a calm demeanor to hide the turbulence inside, I was like an apparently peaceful duck on the top of a lake with its legs rapidly moving to keep it afloat. My kids saw right through the facade. To them, I was like an angry-looking duck going rapid fire.

After one of my tirades, Amy would take the kids somewhere, often to the beach, a place they all loved, to give me time alone to chill. When they came home, Amy would apologize for leaving me alone.

"It's okay," I would say, not telling her that alone was where I wanted to be, though she knew it. Alone, I could stew in rage and frustration. Alone, I could retreat to my man cave and relax playing war video games. Still, being treated like a monster and avoiding me didn't fit my sense of self as a good person. What's more, it made me lonely. I was in a catch-22. While being seen as a monster had the side benefit of being left alone, I still wanted to be part of my family.

Meanwhile, my family was getting the same mixed messages. Do they leave me alone, or should they include me and then risk needing to buy another toaster? They didn't know the answer.

With people outside my family, I was more successful in hiding my anger. When asked how I was coping with civilian life, I'd grit my teeth and say, "I'm fine." A clinical psychologist supervisor once told me that "fine" is just an acronym for "fucked up, insecure, neurotic, and emotional."

While I never considered myself insecure, neurotic, or emotional, the point was well taken: underneath my bluff as being "fine" was one fucked up dude digging himself into a deeper hole of despair and rage.

My misery reached such a boiling point that Amy finally gave me my marching orders, suggesting that we divorce and I go back into the military. I certainly wasn't happy walking around sunny California.

Yes, I thought, *that's exactly what I want—to return to combat.* In the military, I was not obligated to be emotionally connected to anyone.

At the same time, I feared being alone. What if Amy really did leave me, as she suggested? The thought of losing my family was unbearable.

I was lost and needed help. Yet, I didn't seek therapy. That would have forced me to accept that I wasn't "fine."

In spite of all of this, my problems were mild compared to many returning vets who can be ticking time bombs, occasionally beating their wives and children. Some commit violent crimes. Many of the guys I work with would express sentiments like "You can't come up behind a combat veteran or their reaction might be to kill you," or "You should never impede the movement of a veteran because they'll knock you out," and so on. Being treated like a powder keg scratched that itch for them. *I'm GI Joe again*, they thought to themselves. *Watch out, motherfucker, don't mess with me.*

Extinguishing that rage is a challenge. In the military, keeping it meant survival. From the beginning of basic or recruit training, you learn quickly that your only defense in combat is aggression and that you need to suppress any emotion other than rage. As your DI might say, "Let me see your war face! Let me hear your war cry!" Frustrated at your DI because he's in your face? Too bad, stuff it down. Upset that you made a mistake and now the rest of the platoon has to pay for it? Too bad, stuff it down. Worried about the next training event? Better keep that stoic face on for when the DI comes back.

Warriors put on the façade for each other, as well. Some, including me, used the pain of others as fuel. On a forced march, a long foot movement carrying a large amount of weight in a pack, seeing another person "falling out" (not being able to keep up the pace) was like a short turbo charge to me. *I'm better than this man.* Except that "man" is supposed to be your teammate! So don't let your peers know that you've got any emotions other than that hard-core vicious intensity meant for combat, or you'll get left behind.

Even worse, not keeping your cool showed weakness and feminized you. Clay had labeled me "emotional" because I was annoyed when my suggestion to get Marines to Doha for training was denied. That hit below the belt. There was no comeback. If I replied, "I'm *not* emotional," it would have meant I was defensive and *being* emotional. I just had to sit and stew and resist punching the asshole.

Emotions are temporarily rawer sometimes after a fight, as many warriors have to "punch it out" before they can "hug it out." It's the same with alcohol. When affected by alcohol, inhibitions lower, and warriors get emotional briefly before passing out. Afterward, emotions are bottled back up. A tough alpha male stays cool.

Such toxic masculinity is rewarded in the military. Housed in barracks or tents with only men, you get used to being with stressed-out buddies who are on edge. Conversation is often punctuated with "No fucking way, fag," and "What a croc of shit—kiss my balls." Nastiness becomes part of your persona.

Upon retirement, vets are expected to just turn off their training and war mentality and become like everyone else. Often the result is a fake self. In Karl Marlantes' Vietnam War novel *Matterhorn*, two men are returning home from war, and one says to the other, "You'll pretend how sorry you are because you won't be able to explain how good it felt to do things so bad."

Underlying toxic masculinity includes the need to seek and achieve dominance. When vets come home and are ignored and unappreciated, they feel stripped of dignity and significance—the low man on the totem pole. It's little wonder so many lose themselves to drugs and alcohol. Many become pariahs of society.

I'm unsure to what extent I was guilty of toxic masculinity. Beneath my rage and discontent was more a feeling that life was meaningless than the need to suppress "unmanly" emotions.

Figuring out how to return to combat was my only solution to this misery. Seven years after I retired in 2014, during the fall of Kabul, I tried to join the CIA so that I could contribute again. I called my buddies at the agency to volunteer for divisions that I was qualified to serve in. As late as 2023, a SEAL buddy connected me to a naval reserve officer recruiter to see if they could work out a way to "un-retire" me, or at least to bring me back in the reserves.

Though it didn't work out that time, this need persisted. In 2020, I was forty-three years old, the father of four, and working toward my doctorate in clinical psychology. Life should have seemed satisfying. I had been out of the military for five years. And yet I wanted to return.

Even now, as if a little veteran homunculus lives inside me, I consider how I can assist those on active duty. For instance, I could work as a civilian employee for contracting and find Department of Health positions at the nearest Marine base.

In my position at the VA, I feel compelled to go to extra lengths to help another veteran, just as when on active duty, I would have jumped on a grenade to save a buddy. As veterans, we all would have.

Some vets working for the VA supportive housing program are obsessed with trying to find housing for their homeless veterans, despite feeling continuously re-traumatized by their patients' tales of woe. It's as if the little veteran inside says, "Just one more"—to deployments, trips around the block looking for a homeless veteran, or "roger-ing up" to sacrifice established family events to help a veteran in need. Unable to "turn it off," they would keep scanning the streets of Los Angeles for homeless veterans to help, neglecting their own families.

Because of high rates of burnout, several veterans had to discontinue working with homeless veterans.

It's never enough. We are addicted to war and completely committed to our comrades in arms who have served with us in those wars. I'm reminded of a line from George Santayana: "Only the dead have seen the end of war."

In my quest to understand why the military seemed to work for so many, in contrast to civilian life, I find Abraham Maslow's "hierarchy of needs" helpful. Let's now explore one of the most famous and enduring constructs in developmental psychology.

CHAPTER 11

MASLOW'S HIERARCHY

If the essential core of the person is denied or suppressed, he gets sick sometimes in obvious ways, sometimes in subtle ways, sometimes immediately, sometimes later.

—Abraham Maslow

In the last chapter, I talked about my struggles transitioning to civilian life and my desire to return to combat—all symptoms of Warrior Withdrawal Syndrome. What saved me from falling into an abyss—and what gave my life meaning—was becoming a clinical psychologist and helping vets navigate the long road into civilian life.

Maslow's hierarchy of needs helped augment my understanding of the oxymoron that constitutes WWS: being saved by the wish to die.

Devised almost eighty years ago by Abraham Maslow, an American psychologist, the hierarchy of needs is a theory of psychological health predicated on fulfilling innate human needs and remains one of the most enduring and well-known psychological constructs for understanding human development. Beginning from the base of the pyramid, the hierarchy includes physiological needs, safety, love and belonging, self-esteem, and, at the top, self-actualization.

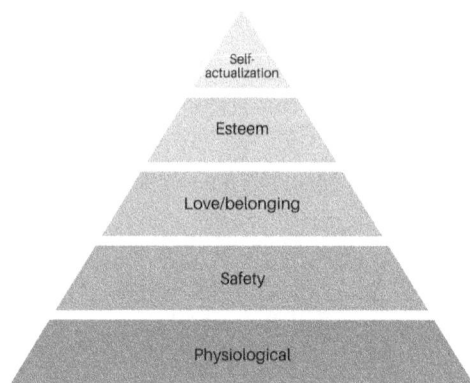

As we go through the hierarchy, we will see how meeting these needs in military life set up a behavioral mindset that doesn't work in civilian life.

Let's look closely at these needs and the contrast between military and in civilian life.

Physiological Needs

Our first and most basic needs relate to **physiology**. These include air, food, drink, shelter, clothing, warmth, sex, and sleep.

The military provides you with some basic needs: clothing, a place to sleep, food, drink, and clean air, at least when you're on base—all for free. If you are married, they pay for an apartment or townhouse.

Once you're deployed though, all basic needs go to hell. The air could be filled with chemicals or flying sand, the temperature is scorching, food comes from a can, shelter is found in a tent, sleep is taken in a hammock, and clothing amounts to a filthy uniform. And, of course, no sex.

In these respects, home might seem to have an edge in uniformly providing clean air, nourishing food, safe shelter, comfortable clothing, warmth, sex, and sound sleep.

Yet, this isn't always the case. In fact, a significant number of returning vets become homeless, can't afford proper food or new clothing, and sleep fitfully, especially those with PTSD. Some feel compelled to return to the military just to be assured of food, shelter, medical care, and so on. So, yes, home might have an edge on physiological needs for some, but not for all returning vets.

Added up, both military and civilian lives may or may not provide basic needs. They can be found in the military, but not when deployed; they can be found in civilian life, but not for those lacking basic skills who can't find gainful employment or those who come home with physical or emotional wounds that prevent a basic return to society.

Safety Needs

Once physiological needs are satisfied, the next need to ensure survival is **safety**: order, predictability, and control in your life.

Safety in the military is a double-edged sword. On the one hand, the military offers you safety by guaranteeing you employment, social welfare, law and order, financial security, and, if you choose to make service your career, future stability.

None of these things are certain at home. This lack of guarantee is one of the major obstacles in adjusting to civilian life that leads to WWS. Many of the retired vets I work with have been unable to find gainful employment or have gone from dead-end job to dead-end job.

At the same time, the "safety" of being provided for in the military comes with the constant danger of losing your life. Though I was excited to be in the thick of things in Iraq, an element of fear also surged inside me when outside the wire. At home, where I lived without the risk of death, I was able to go about daily business without constantly looking over my shoulder. On the other hand, those with PTSD, which I did not suffer from, carry this fear with them into civilian life.

Another element of safety involves order and predictability. By controlling all aspects of life, the military shines in this regard, in contrast to the chaos and instability that often greets returning vets. Yet, while this is mostly a plus for those looking for that kind of discipline, it can also be a minus, because the self-control you normally enjoy in life is sacrificed to the authority of higher-ups. Rules must be followed. There is no wiggle room. Contrast this to civilian life, where you often have the option of being your own boss for work.

Love and Belonging Needs

Once physiological and safety needs have been met, the next needs for survival in the hierarchy are **love and belonging**. Humans are social creatures, and connection and belonging are paramount to psychic survival. Examples of belonging needs include friendship, intimacy, trust, acceptance, receiving and giving affection, and the commitment and loyalty that come with love.

I had all of the above from family and civilian friends. Yet, I preferred spending time with my military buddies who, after I retired, became my primary circle of friends.

There is no bond quite like the one you share with your band of brothers. It is one of the top reasons vets long to return to combat. I'll discuss this in detail in chapter 7.

Your relationship with your senior officers is another story. As I experienced, you can be stuck with higher commands who are inept and sometimes even dangerous, but you have no choice except to obey. Unlike civilian life, you can't quit your job in the military if you don't like your boss.

Of course, your military buddies don't provide the more intimate or romantic love, affection, and sex you can get in civilian life. Even so, civilian relationships for many returning vets lack the shared experience of military life or become too fractured to have these more intimate benefits, and too often, the vet's only desire is to return to his band of brothers.

In spite of this, other relationships can be repaired and become loving and satisfactory. Although I came home to a loving family and an exceptionally understanding and supportive wife, I was too filled with rage to benefit from their love. It took time before I was able to fully embrace them and time for them to fully embrace me.

Esteem Needs

Once the first three needs have been satisfied, the fourth need in the hierarchy involves **esteem**, including self-worth, accomplishment, and respect. Maslow classifies esteem needs into two categories: esteem for oneself (reached through dignity, achievement, mastery, and independence) and the desire for reputation or respect from others (exhibited through status and prestige, among other expressions of admiration and validation).

Building Skills

The military offers numerous opportunities to earn esteem, to stretch beyond what you thought you were capable of. Though I had been athletic, my training as a Marine took me to new heights. I could do one hundred crunches in two minutes, run three miles in eighteen minutes, and do as many pull-ups as anyone else.

I learned how to jump out of airplanes, navigate obstacle courses on my belly, and hold my breath underwater longer than I imagined I could. By doing this, I built resilience and self-confidence in my abilities. All the while, my buddies cheered me on as I executed nearly impossible feats.

Diminished Independence

At the same time, the military does much to erode self-esteem by diminishing independence. From the get-go, the military encourages dependency.

Recruits are young, between the ages of seventeen and twenty-five. At these ages, you're still establishing your identity. When you enter basic or recruit training, you will establish the collective identity of a warrior, not an individual, unique one.

Obedience to higher-ups is the call to duty. Personal choices, for the most part, are off the table. All needs are managed by the military, including your income, what you wear, what you do, what you eat, where you sleep, and who your peers are. You are no longer an individual, but instead an extension of Uncle Sam.

To further diminish identity, DIs yell at you from all sides, and you're helpless to do anything more than say, "Yes, sir," and "No, sir." While this seems harsh, it's necessary in order to get the civilians conditioned and transform the recruit into a good trooper.

The list of dos and don'ts is long. No longer can you put your hands in your pockets, because that will draw the ire of a DI. You lace all your boots "left over right" because if you don't, the DI is going to make you do more push-ups.

Yes, you're a "motherfucker," but only with your peers; with your superiors, you're an underling. Mouthing off or disobeying will get you thrown in the brig. Your job is to be in line with what the military wants you to be and not what you used to be. What matters is the service record book, the DD 214 (your discharge papers), the medals and ribbons, etc. That's what defines you, not who you are individually.

Self-Actualization

The fifth and highest level in Maslow's hierarchy is the need for **self-actualization**. This level of need refers to realizing personal potential, self-fulfillment, seeking personal growth, and peak experiences. In Maslow's own words, "to become everything one is capable of becoming."[1]

Purpose

To work toward self-actualization, you must have a sense of purpose in your life. For many, the military provides this, especially for those who were limited in what they could do and achieve before entering the military. Serving your country, keeping your fellow citizens safe, gives meaning to your life and offers the opportunity for growth and self-actualization.

Once you leave the military, your sense of purpose and drive might be lost unless you find a satisfying substitute, like joining a police force or the FBI, which is what I did briefly. If not, you wander in darkness and do whatever it takes to pass the time, all the while wanting to return to combat.

For this reason, I work hard to help vets find a sense of purpose in their lives. Until they do, they flounder, and their growth gets stymied.

Autonomy

Another important component of achieving self-actualization—becoming a fully realized, mature human being—is autonomy. Autonomy is the need to perceive that you have choices, that your actions result from your own volition.

In the military, autonomy is discouraged; instead, conformity and compliance are prized. You know your place in the pecking order and follow orders, whether you agree with them or not.

Roles are strictly assigned, and hierarchy must be respected. At the U.S. Naval Academy, the only five authorized responses to a senior officer or midshipmen are "Yes, sir," "No, sir," "I'll find out, sir," "No excuse, sir," and "Aye, aye, sir."

From the start of initial entry training, DIs work hard to strip you of individuality and turn you into a member of the machine. Hostility is maintained through competition to train and foster "the warrior spirit," as well as through insults and demeaning comments. You can do nothing to stick up for yourself unless you want to clean the latrine floor with a toothbrush.

When you are assigned a regiment, the intimidation continues from your commanders. If your boss is a jerk—and I had plenty—tough luck. In addition to Marone and Clay, I had senior officers hold up my orders, reports, or even requests to be with my family until I modified my behavior to fit their model.

Another form of intimidation is "monitoring," or "digital clocking." While there isn't a punch card per se in the military, an ethos exists that you have to arrive early and leave late. You feel as if Big Brother is always watching what you're doing, and evaluations of your behavior are plentiful. The higher headquarters keep tabs on things like the following: how long someone has been deployed, what their overseas control date is, how many recruits they have successfully processed, and more. All of this would factor into a subjective formula to determine things like promotions and assignments.

If you haven't hit the "right wickets," then presumably you're not eligible for promotion. If your scores are too low, you will have to work on improving them while also relying on seniority to improve your opportunities for promotion and assignment.

These and other destructive components of the military mess with your identity as an autonomous, competent individual. Depression sets in, and vets might count the days till it's over. Then they go home to discover that they've lost their sense of self. This is one of the experiences I typically start with in a WWS therapy group, and I will go into detail about it in Part V.

In the next section dealing with identity, we will spend time explaining more of these hierarchical needs.

PART IV

COMBAT ADDICTION

CHAPTER 12

WARRIOR WITHDRAWAL SYNDROME: CONSTELLATION OF SYMPTOMS

Civilians balk at recognizing that one of the most traumatic things about combat is having to give it up.

—Sebastian Junger, *War*

After years of much thought, discussions with mental health practitioners, and research, I devised a theory to explain the symptoms of Warrior Withdrawal Syndrome.

Initially, I had named it BAMF ("bad ass motherfucker"). I chose this acronym because it was a title, not a label. This made it more comfortable for vets to wear because it lacked the stigma attached to being diagnosed with a mental health disorder. Many do not want to be labeled with PTSD or any other psychiatric disorder.

I changed this to Warrior Withdrawal Syndrome because I feel this best describes the condition.

In figuring out the symptoms, I began with the premise that a person's baseline functioning—their status quo and where they normally operate—would be modified by military service.

After four years or more in the military, a warrior identity is deeply entrenched in your psyche. Once you get home, these changes in identity are not easily undone. One vet I worked with still laid out his equipment before going for a hike to be prepared "for anything." Another checked corners for enemies and sat with his back to the wall in self-protection.

It takes time to retrain and return to a semblance of your previous baseline of functioning prior to service. It also takes support from families, decent employment and health services, and for many, adequate mental health services. Unfortunately, there are long delays in securing psychiatric appointments. The VA's common response to a PTSD diagnosis is simply to ship a packet of meds to the patient. Left to their own devices in titrating medications, some vets become addicted.

Defining WWS

WWS is neither a diagnosis nor a disorder. No studies exist to determine if the cluster of symptoms, as previously defined, require a distinction in the latest edition of the *Diagnostic and Statistical Manual of Mental Disorders* (DSM-5).

Nor is WWS the same as post-traumatic stress disorder (PTSD), which many vets suffer from. There is comorbidity: in other words, many vets with PTSD also have WWS. I've had a few veterans come through my clinic who'd seen significant combat action—everything from having to pick up the pieces of fellow Marines following explosions to the buddy right next to them being blown up "instead of me." These were specific traumatic events that continue to elicit the textbook PTSD responses of intrusive memories, negative thoughts, avoidance, changes in mood, and strong physical and emotional reactions.

At the same time, many with WWS don't have PTSD. They don't suffer nightmares, endure panic attacks, or jump when a child screams.

These differences convinced me that WWS was a separate disorder from PTSD. There does exist a subset of PTSD in the DSM-5 called "combat addiction," which describes returning vets who are addicted to war. All who were diagnosed with this subset had experienced heavy combat. Further on, I'll describe this syndrome.

Let's first look closely at the symptoms of PTSD compared to WWS.

Comparing Criteria for PTSD Diagnosis with WWS

There are five criteria for a PTSD diagnosis:

1. Exposure to a traumatic event
2. Presence of "intrusion" symptoms related to the traumatic event
3. Avoidance
4. Negative changes in thinking and mood
5. Changes in reactivity associated with the traumatic event

Exposure to a Traumatic Event

This criterion involves exposure to actual or threatened death, serious injury, or sexual violence in one or more of the following ways: directly experiencing this as a witness, learning of violent or accidental trauma to family members, or prolonged exposure to details of traumatic events. These all constitute exposure to a traumatic event.

While traumatic events like combat can exacerbate WWS symptoms—including a wish to return to combat, poor post-military adjustment, and difficulty controlling anger—diagnostically, WWS symptoms can manifest without a significant or specific trauma, as in my case. In fact, even those who never saw combat experience the symptoms.

Presence of "Intrusion" Symptoms Related to the Traumatic Event

"Intrusion" symptoms related to the traumatic event include things like flashbacks, dreams, recurring distressing memories, etc. One vet I worked with would wake up thrashing about in a panic because he couldn't find his weapon. Other veterans have described nightmares with a persistent fear of being attacked or chased.

With a WWS experience, no intrusion symptoms need to be present because there doesn't need to be a traumatic event. Again, many vets I work with have never even seen combat.

Avoidance

A person with PTSD will make efforts to avoid memories and external reminders of their traumatic event. In contrast, a veteran with WWS will often actively seek out other veterans, continuously contemplate returning to active duty, and counterintuitively find comfort in the familiar pain and suffering of deployed life.

Negative Changes in Thinking and Mood

Negative changes in thoughts and mood associated with the trauma can take different forms:

- Not being able to remember details of the trauma
- Negative thoughts about oneself
- Distorted thoughts about the cause of the trauma and who should be blamed
- A persistent negative emotional state
- Feelings of detachment
- Diminished interest in activities and participation
- Difficulty experiencing positive emotions

With this criterion, overlap exists between PTSD and WWS. Negative thoughts about oneself, a persistent negative state, detachment from civilians and family, and challenges with feeling positive emotions are also present with a WWS veteran, as I detailed in chapter 8.

Negative emotions had survival value in the military and are hard to extinguish when you get home, especially anger. When deployed to a combat zone—especially if you went outside the wire—anger was your best friend. It drove you, fueled you, and kept you awake and sharp.

Likewise, it's difficult to experience happiness or joy when you can't get a job and find yourself constantly going to the funerals of buddies who were killed in combat—or killed by their own hand, because they, too, struggled with making it in the civilian world.

In a 2022 online article from *Psychology Today* titled "Why Veterans Feel Addicted to Combat," soldier David Kendrick wrote of his personal experience with WWS. Unable to adjust to civilian life, he was obsessed with returning to combat. Lacking the normal range of human emotions, he was indifferent toward his own suicide attempts, partly because he was emotionally separated from his family and numb to their emotions. "There would be times when I would look at my own mother and feel nothing," he explains. He felt dead inside because the only thing that he'd known from age eighteen to twenty-three was "kill the enemy." Unable to adjust to civilian life, he tried to reenlist even though he had fought hard to get out of the Army.[ii]

Changes in Reactivity Associated with the Traumatic Event

Changes in reactivity associated with the traumatic event are evident from a person's angry outbursts, reckless or self-destructive behavior, hypervigilance, exaggerated startle response, problems with concentration, and difficulties sleeping.

With WWS, there may not have been a traumatic event, as noted. The vet will show angry outbursts, reckless or self-destructive behavior, problems with concentration, and difficulty sleeping—all because of a loss of identity and purpose when transitioning to civilian life.

Likewise, they will continue to show vigilance because it's become hardwired into their brains. Neuroscience has a saying: "Neurons that fire together, wire together." Four years of repeated behavior, thousands of hours on ranges or "outside the wire" where danger lurked around every corner, hundreds of hours scanning for IEDs established a pattern of hypervigilance and transformed the vet's brain into a warrior mentality. And the service encourages it because they need their troops to jump and respond without pause to explosions and enemy fire.

Detailed awareness of your surroundings prepared you for a potential assault and kept you safe. Warriors who noticed the trip wire would survive, while those who didn't pay attention would be killed.

One Marine veteran of heavy combat in Afghanistan said, "The PTSD you come home with is everything you had to be to survive in combat; hyperaware of every sight and sound until you become a total paranoid, not being able to sleep for shit, and ready to waste anyone that gives you the wrong vibe. You had to get PTSD to stay alive."[2]

When you get home, you carry these emotions with you. Preparing for some untoward, dangerous event to happen is hard to extinguish; the anger and vigilance remain. How do you learn to turn off the need to scan the rooftops and check the security of the perimeter whenever you hear a noise in the middle of the night?

At the same time, those vets with WWS but not PTSD are not showing the extreme hypervigilance that defines PTSD. I will sit spontaneously in the seat in the restaurant with the best tactical vantage point, even if I don't expect anyone to come in with an AK-47 rifle. I do it "just in case." At the same time, I don't tremble if a dark-haired, mustached Arab saunters past my table with a Fedayeen tattoo down his arm and what I imagine is an evil eye.

Combat Addiction: Subset of PTSD

As demonstrated, there are clear distinctions between PTSD and WWS. There is, however, a subset of PTSD that captures the experience of a significant number of those who suffer WWS.

In the article "Combat Addiction: Revisited and Reaffirmed," the authors identify combat addiction (CA) within the fourth cluster of post-traumatic stress disorder (PTSD) in the DSM-5.[3] In those with CA, combat was experienced as a "high." To replicate this adrenaline rush, they engaged in aggressive, reckless, or self-destructive behavior, like dangerous driving and substance abuse, in civilian life.

Among veterans of heavy combat, the authors state, CA is present in the vast majority of those exposed. The condition appears to persist for well over a decade, with only very advanced age suppressing it. In my own practice, I've found this to be true of Vietnam veterans I've treated who are now in their seventies.

CA was first formally described by L. P. Solursh, a psychiatrist treating Vietnam veterans, and it is recognized as a specific and treatment-resistant form of PTSD. It manifested most clearly in those who experienced intense, repeated combat exposure. Interviewing a sample of one hundred veterans of active combat experience, Solursh reported that 94 percent recall battle as exciting, empowering, and inducing "a high," with the majority compelled to keep loaded guns at the ready and reenact combat scenarios in outdoor environments.

Being sensation seekers, the soldiers wished to recreate the thoughts, feelings, and actions derived from combat experience to reconstruct a parallel state to the original trauma of living on edge with adrenaline flowing. They enjoy war and find it hard to fit into a civilized world.[4]

In addition to aggressive behavior, they seek prescribed or recreational stimulants. In the Vietnam veterans, chronic alcohol dependence was nearly unanimous.

Researchers in 2008 found that those veterans with greater exposure to combat involving killing and severe wounding had the greatest propensity for risky behaviors upon retirement, such as reckless driving and other high-thrill activities.[5]

In a later study, researchers sampled over two thousand combat soldiers within three months upon returning from a year of wartime duty in Iraq in 2006.[6] Results showed scores significantly differentiated heavy drinkers, dangerous driving, and subjects reporting recent aggressive outbursts.

In a study of two Danish units deployed to Afghanistan, it was found that those deployed were strongly motivated by excitement after their return, even when controlling for gender and age.[7] The study concluded that whether or not one deployed to Afghanistan is not what determined increased excitement motivation; rather, it was whether or not a subject had engaged in combat. The author concluded that combat soldiers can become adrenaline junkies because their physiological threshold for excitement is altered by the exposure to such extreme danger.

In my practice, I've found that, of those who experienced combat, the group most likely to show stronger withdrawal symptoms at retirement are the special forces. These guys are stronger, more determined to succeed, and share a positive attitude toward punishment, which enables them to pass the difficult courses required to qualify for special forces. As part of training, they were told they weren't going to make it—there was no way. Their response? "Watch me."

To understand what neurophysiological factors might underlie combat addiction, Solursh references a 1985 study that explored a possible addiction to trauma.[8] This earlier study postulated that reexposure to trauma may produce a paradoxical sense of calm and control due to opioid release. Cessation of the traumatic stimulation is followed by symptoms of opioid withdrawal. This process could account for a veteran's voluntary reexposure to trauma, pursued as an attempt to master the meaning of the trauma.

Distinguishing WWS from CA

While combat addiction might sound the same as WWS, there are significant differences.

Never in Combat

To begin with, I've treated vets with the desire to reenlist who have never witnessed combat, and they do not fit the profile for CA. Their reason for wishing to return to service was a preference for the simplicity of their lives while on active duty. They understood their purpose and had a seemingly firm grasp on their identity.

Never deployed to combat, the untested veterans experienced a correlative feeling to combat addiction: combat envy. Even among those who did go to combat, they may still experience combat envy when they're out of the action.

Not Sensation Seekers

Secondly, not all who wish to reenlist are sensation seekers. Being a classic case of WWS, I serve as an example. To this day, I neither engage in reckless behavior nor consume alcohol or recreational drugs. Generally, I don't like risk. I don't like biking, rollerblading, or skiing at a high rate of speed, as these things seem like they could get out of control quickly.

Outside of the military, I live a subdued life. I prefer reading or playing a videogame to riding a motorcycle or parachuting. Even while on active duty, I canceled enough parachuting opportunities that the Army threatened to rescind my incentive pay as a paratrooper. I do enjoy rollercoasters, as long as they're not the rickety ones found at local carnivals with a bunch of loose screws and duct-taped repairs.

Needing a High Is Only One Motivation to Reenlist

The desire to reenlist involves more than just a desire to experience a high. It includes the need for fame, a tribe, a purpose, and a strong masculine identity. For these reasons, CA is a subset of WWS, not the entire picture.

Substance Abuse Disorder

Aside from the overlap with the CA subset, WWS differs considerably from PTSD, but it includes symptoms eerily similar to the constellation of symptoms for substance abuse disorder (SAD) in the DSM-5. The only difference is that, rather than being addicted to a material substance, a person with WWS is addicted to service, combat, or the idea that they are a warrior.

When I discuss WWS with veterans, I don't try to label the "substance" for them. It differs from veteran to veteran. In my case, I might need external validation that I'm tough or a warrior. Even now, I struggle not to throw my military experience into a conversation, and I'm embarrassed, for example, to tell people how long it took my fourteen-year-old son and I to finish a marathon (just under six hours), even though I was still recuperating from knee surgery.

The eleven criteria for SAD are the following:

1. Taking the substance in larger amounts or for longer than you're meant to

2. Wanting to cut down or stop using the substance but not managing to

3. Spending a lot of time getting, using, or recovering from use of the substance

4. Cravings and urges to use the substance

5. Not managing to do what you should at work, home, or school because of substance use

6. Continuing to use even when it causes problems in relationships

7. Giving up important social, occupational, or recreational activities because of substance use

8. Using substances again and again even when it puts you in danger

9. Continuing to use even when you know you have a physical or psychological problem that could have been caused or made worse by the substance

10. Needing more of the substance to get the effect you want (i.e., tolerance)

11. Development of withdrawal symptoms, which can be relieved by taking more of the substance

Let's compare these criteria to WWS, one by one.

Taking the Substance in Larger Amounts or for Longer than You're Meant To

While many vets adjust to civilian life, a small number, like me, continue to deploy past the point they're meant to serve, and they seek more and more combat or higher and higher ranks. Some don't stop serving until they are wounded and unable to serve anymore. Others only stop serving because they reach their service limit by age or rank and don't qualify for continued service.

Wanting to Cut Down or Stop Using the Substance but Not Managing To

At one point, I thought I had had enough of the military and tried to cut down on how much I was serving. I made some efforts to get out in 2003 and again in 2012, when I was told JSOC would "dead-end" me. Yet, despite my best efforts, I continued to try to return to active duty years after my service ended, my latest attempts being to apply to the Medical Service Corps in 2020 and to the Navy Reserve in 2023.

I've had to struggle to maintain boundaries between myself and the vets I've worked with. For instance, I had to control my desire to follow their paths once therapy was terminated. Other veteran VA employees I know suffer the same struggle. They'll spend more time at work than with their family because they can help one more veteran. Meanwhile, their son keeps glancing up at the stands during his baseball game, looking for his warrior father.

I've served with and treated numerous veterans who returned to active duty with "broken time" because some event pushed them over the threshold to jump back in. For many, it was 9/11.

More recently, the deciding event was the rushed and chaotic withdrawal from Afghanistan in 2021, including the suicide bomber attack that took the lives of thirteen service members. I put in a call to the CIA to work for them after that incident, telling myself that I'd eventually return to my studies and complete my degree. Fortunately, I recovered somewhat from this bout of WWS and didn't follow through with a complete application, though the cravings and urges to go back are still there.

Spending a Lot of Time Getting, Using, or Recovering from Use of the Substance

Basically, from the time of my retirement, I put my energy into both trying to redeploy and trying to effect boundaries from military involvement.

I speak to my JSOC buddies more than my own brothers. I bought tickets to visit a JSOC buddy at Walter Reed Hospital as soon as I could, whereas I was too late to do the same thing for my own brother before he passed away from cancer.

At one point, a selection board that could have been a gateway for me to more deployments challenged my priorities. I listed "family" above "service." The board cited how little time I actually spent with my wife and kids, and how I was now at the board *asking to spend less time with them by deploying again!*

Most of the active-duty Marines I served with similarly prefer to spend more time with military buddies than with family.

Cravings and Urges to Use the Substance

Whether for the pursuit of honor, valor, a high, a tribe, glory, or tiny bits of ribbon that convey prestige, an inextinguishable fire burns inside many veterans for combat. This fire burns through marriages, relationships, and post-service job opportunities as the veteran quenches the fire with substituted drugs, alcohol, unhealthy obsessions, or risky behavior.

I have witnessed veterans conceal their physical injuries in order to deploy. I have seen them attempting to hide mental health issues, sometimes paying cash to civilian providers to avoid any record of therapy. I've been consulted on photos that were digitally manipulated to make it seem like a veteran was following regulations.

Some veterans I've worked with have destroyed letters that would exempt them from service, and others have fabricated material to be able to extend their service.

Not Managing to Do What You Should at Work, Home, or School Because of the Attachment to the Service

As I've described extensively, I neglected my family for a chance at combat, including choosing to live away from them for more than three years.

When charged with my kid's care, I did the bare minimum so I could play video games, the one thing that distracted me from my obsession with returning to combat.

Continuing to Use Even When It Causes Problems in Relationships

While on active duty, the service member places more weight on their professional relationships, setting aside their family and social relationships. This causes problems with spouses, children, and other family members, especially while constantly pursuing deployments. Divorce rates among veterans are uncommonly high and are even higher in the special operations community, with recent statistics as high as 90 percent for SOF communities like the Navy SEALs.

In the past year, the list of divorced buddies of mine continues to climb. A guy I deployed with unexpectedly dropped divorce papers on his wife of almost thirty years. Some of these guys are now on their third and fourth marriages. Others simply swear off women.

Giving Up Important Social, Occupational, or Recreational Activities Because of Substance Use

Many vets I know will choose to deploy or spend time with military buddies and miss a significant family event, thereby communicating to their partners and children that they are undervalued.

The vet will delay getting home to their spouse and kids because of a work obligation or extended deployments. Some will volunteer for deployments and then tell their family they were ordered to do it. When I was the officer of the day, I fielded calls from spouses and family members asking if their loved ones were really "in the field" and training rather than simply avoiding home.

While I was with JSOC, I had a soldier who could set his own deployment schedule. In JSOC, you're supposed to be limited to thirty-day trips because the stress of staying longer could cause mental health issues. This soldier was scheduling four-month deployments that covered the holiday period, so he'd miss Thanksgiving, Christmas, and New Year's Eve, apparently wishing to be away from his wife and kids during these holidays. The value of his trips wasn't advancing the mission so significantly that it required that much time from him. He was simply avoiding his family.

Using a Substance Again and Again Even When It Puts You in Danger

Deployments put the veteran in risk of getting maimed or dying.

Even training can be dangerous. Recent catastrophic training accidents have claimed the lives of Marines, from an AAV sinking to paratroopers having trouble opening parachutes.

Yet, we keep putting our hand up—more combat patrols, more time outside the wire, all in search of quenching that need the service instills.

This leads to the ninth criterion.

Continuing to Use Even When You Know You Have a Physical or Psychological Problem That Could Have Been Caused or Made Worse by the Substance

Despite common physical issues, like leg and back injuries from parachuting, warriors keep jumping out of aircraft.

Numerous warriors suffered significant injuries while deployed and still did their best to return for more deployments. In 2003, I had a Marine who lost his eye while we were training in preparation for Iraq. He was shipped home and missed the invasion. Even with only one eye, he went on the next deployment with 2/1 that ended up with major combat action in Fallujah, where many soldiers and Marines died. He continued to serve past that as a DI at the recruit depot in San Diego.

In 2006, one of my Navy corpsmen was seriously injured in an IED strike. To stay on active duty, he fought the medical board that would have retired him due to his injuries. His leg and hip were significantly damaged, making it hard for him to walk, much less run. He was determined to rehab his leg and eventually be able to meet the minimum fitness standards. He ended up making the next deployment and continued to serve in the Navy.

Psychological problems may be more elusive. At JSOC, we had a sergeant major (E-9, the highest enlisted rank) suffering from significant PTSD. He kept deploying because that was what was expected of him and what he expected of himself. Though he tried to keep it together, he degraded slowly from alcohol abuse. Still, he continued to deploy. On his last deployment, he engaged in some reckless behavior (in the category of "loose lips sink ships") that compromised his position and put his unit in danger. The fallout from that cost him his career and set a fire in him to talk to the rest of the JSOC community about PTSD.

Needing More of the Substance to Get the Effect You Want

Each tour and deployment you safely come home from builds tolerance, and with that tolerance, you redeploy over and over again.

Young Marines I taught at SOI would deliberately fail their final exams if the unit they were being sent to wasn't going to deploy to combat. This way, they could recycle and get another shot at orders to a unit that would likely face combat.

Other veterans transition to the reserves to continue to serve while pursuing a civilian career. I've worked with many reservists who, when the Global War on Terror started, volunteered to be activated, almost to a person, so that they could contribute.

In the Danish article I previously cited, the authors suggest that, just as real drug addicts build up a physiological tolerance to narcotics, soldiers can become "adrenaline junkies," because their tolerance toward excitement is "pushed upward" by being exposed to danger.[9]

One combat-addicted veteran started crushing and snorting his oxycodone HCI to achieve the euphoric state akin to the aftermath of a firefight.

Development of Withdrawal Symptoms, Which Can Be Relieved by Taking More of the Substance

When things don't go well, the vet in civilian life regresses back to what worked while on active duty: substance abuse, violence, denial of issues, anger and hostility toward others, aggressive barking commands, suicide, and homicide.

Driving fast on the highway or getting into bar fights makes them feel like they are back on active duty. In the 2017 article "Combat Addiction: Revisited and Reaffirmed," the authors describe a veteran who had been a fully armed and battle-engaged medic with an airborne unit. Upon retirement, he became certified as an emergency medical technician but was unable to separate his civilian function from combat experience. During a "rescue," he beat an injured man with his fist because the man had caused a car crash after trying to abduct a woman he had assaulted.

Discharged from his job, the former EMT began to drink heavily and patrol the city streets seeking confrontations with "anyone Arab." At one point, he saw a man he assumed was of Middle Eastern descent arguing with a white woman in a parking lot. He followed the couple into their building, and when he saw the man put his hands on the woman, the veteran attacked him, beat him, and then fled the scene.[10]

In a 2011 study of returning veterans, anger and alienation predicted increased risk-taking behaviors, like driving recklessly and carrying an unneeded weapon.[11]

Having looked at the addictive symptoms of WWS, let's now explore its dynamics.

CHAPTER 13

SEEKING MACHISMO

It was the Marines who taught me how to act. After that, pretending to be rough wasn't so hard.

—Lee Marvin

One day, I was coaching my son's soccer team. I looked on helplessly as an active-duty Marine lambasted his eight-year-old kid for his poor performance. The kid only wanted to please his dad and get his attention. But the dad wanted another Marine, an alpha kid. Instead, he got a beta crybaby.

There's no crying in the military, as the saying goes.

Military men are badass motherfuckers. They embody machismo.

Machismo is part of the Marines code of conduct. According to *Encyclopedia Britannica*, it refers to an "exaggerated pride in masculinity, perceived as power, often coupled with a *minimal sense of responsibility and disregard of consequences*" (my italics). As a Marine, you value characteristics culturally associated with the masculine and denigrate characteristics associated with the feminine. When Clay described me as "emotional," for example, that hit me in the gut, because it made me sound feminine.

To not appear weak in the eyes of their seniors and peers, Marines will go to extraordinary lengths to conceal pain, fatigue, hunger, thirst, and injuries. Acting as "bullet sponges," they will volunteer to be at the tip of the spear—in other words, to be the first to go into battle.

Historically, being a warrior was a common rite of passage into being a man. Ancient Greek philosophers were also generals. Feudal kings drew their armies from the farmers' fields in their fiefdoms. Even now in Iraq, every male-led household includes an AK-47. It is only recently that we have not equated being a warrior with being a man. This helps to explain why, until recently, women were excluded from service.

The "warrior" version of masculinity pervades Western mass culture. Guns figure prominently in literature, movies, and video games aimed towards a predominantly adolescent male audience. Potentially interpreted as a phallic symbol, the gun is a common component of boys' toys. If boys aren't given toy guns to play with, they'll often make them for themselves. During military training, firing your gun can give you a wild thrill. One retired veteran obtained a concealed-carry pistol permit and routinely wandered the city streets late at night, hoping someone would mess with him.

Returning vets often wear their military uniforms, keep the high and tight crew cut of the military, drive SUVs or muscle cars, and tattoo their bodies with Marine Corps "moto tats," or motivational tattoos, often an eagle, globe and anchor, or bulldog head. They broadcast their service to the civilian population with medals and pins on their lapels and stickers on their vehicles.

Being around bombs exploding, random sniper fire, and war casualties fueled your brain with dopamine and filled your body with testosterone. As a man, it felt normal, because you were in an environment where masculinity provided an edge and was commonly celebrated.

How much did machismo have to do with my own WWS? I'm not sure.

Upon graduating high school, I applied to Boston University, because they offered me an academic scholarship based on my high SAT scores and my need for financial assistance. Even with this help, I still wasn't able to cover room and board. Because of this, once I started at the university, I applied for and got a full scholarship from the Navy ROTC program, with the plan of going through the Navy and ultimately becoming a doctor like my father.

Somewhere in my first year in pre-med, I felt I didn't fit the model. I switched from pre-med to a combat specialty that would allow me to train to become a Navy SEAL.

I didn't know anything about special operations, including the difference between a SEAL and a special forces soldier. Nor did I have preference for the SEALs over Army special forces or Air Force Pararescue. SEALs tend to have massive egos, and I wasn't necessarily looking to add to mine. All I dreamed of was becoming some kind of commando, or real-life G.I. Joe, like Arnold Schwarzenegger's character from *Commando* or *Predator*. When you grow up with three brothers, there's a lot of competition to be the strongest and fiercest.

Generally a service agnostic, I liked Amypolis and the Navy more than the Army. When I decided to stop pursuing medicine, I already had a Navy contract, in part because in my freshman year, the Navy could guarantee me full tuition, while the Army could not. With the Army, I would have had to rely on the school to cover the difference. Nonetheless, I interviewed with both.

I applied to be a Navy SEAL because I fit the MO. The characteristics of Navy SEALs include intelligence, physical fitness, and key personality traits such as honor, commitment, and courage.

I was a good, smart student, and I liked to think of myself as honorable.

As for commitment, the SEALs have a 40-percent rule. When you feel exhausted, cooked, and totally tapped out, you're really only 40 percent done—you still have 60 percent left in your tank. That was me. I had machismo right from the get-go, and I would keep going. I needed to be the fastest, strongest, and most determined, or I would get frustrated. I committed to being the best. To toughen me up, trainers would tell me, "You're not going to make it. There's no way." Just like other SEAL candidates, I would need to respond, "Watch me."

Courage comes down to being able to proceed even when you are afraid. Constitutionally, I'm a calm guy, and it takes a lot to ruffle my feathers.

As for physical prowess, I'm a definite "10." In high school, I was in varsity sports on the lacrosse, football, and wrestling teams. In college, I continued to play lacrosse and rugby before deciding to try out for the wrestling team.

I felt confident that I could fulfill the physical fitness test requirements to be a SEAL. They were the following:

- Swim five hundred yards in 12.5 minutes.
- Do forty-two pushups in two minutes.
- Do fifty sit-ups in two minutes.
- Complete six pull-ups.
- Run 1.5 miles in eleven minutes.

In the end, my physical prowess or SEAL-like qualities didn't matter, because in order to qualify to become a SEAL, you need excellent 20/20 vision. Mine was something like 20/2000! In fourth grade, I got my first pair of glasses because I couldn't read the chalkboard. I refused to wear them, thinking they would peg me as a "sissy." Needing to prove I was "all boy," I learned how to memorize what the teacher said in an attempt to get by without glasses. Within my first year of college, I knew I was disqualified from the SEALs because of my poor vision.

Ultimately, I came to realize that the SEALs would not have been the best choice for me. At age four, my older brother pushed me into a pool, and I wasn't able to swim. Luckily, the adults rescued me, but afterward, I was afraid of drowning. With much training, I was able to swim reasonably well and felt more comfortable in the water than most. However, I also learned while training in Coronado that I'm negatively buoyant, which makes surface swimming a lot harder for me.

If I couldn't be a Navy SEAL, I chose the next closest macho special forces: Marine Force Reconnaissance (FORECON). FORECON is one of the Marine Corps special operations that supplies military intelligence to the Marine air-ground task force. I liked the fact that the recruits had to be more physically fit. We were required to jump out of planes and scuba dive. We were also involved in direct action, shooting at bad guys. During large-scale operations, I'd be involved in direct action and deep reconnaissance. For instance, I would assess battle damage, or set up and prepare LZs and drop zones for operations.

How much did machismo have to do with my desire to join the military and go back to combat? Again, I'm not sure. I do know that being denied medals—those all-important symbols of heroism—in 2003 was a blow to my masculinity and a setback to my need for validation and the recognition of being a bad ass motherfucker.

A bed partner to the drive for machismo is the drive for fame, which I will explore in the next chapter.

CHAPTER 14

SEEKING FAME

A soldier will fight long and hard for a bit of coloured ribbon.

—Napoleon Bonaparte

Warriors want fame and recognition for their service—if not from the civilian population, then at least from each other. They want to feel they are heroes. Being one is measured visibly by the medals attached to your lapel, especially those that show you were in war. War can make you a hot commodity. One action in Iraq or Afghanistan can make up for a career of ineptitude. It can promote you far beyond your actual capability because we generally worship our war heroes.

In the Marines, you hope to earn the coveted CAR, an external validation that brands you as a courageous warrior. It is so prestigious that Marines who have earned it buy massive decals of the CAR to put on the rear windows of their vehicles. Some had the CAR tattooed on their bodies. Hats, T-shirts, and other accessories with the CAR became ubiquitous around the Marine infantry camps. Items with the CAR design would sell out as soon as one of the Marine units returned from deployment.

The same was true of the Combat Infantry Badge (CIB) and, to some extent, the Combat Action Badge (CAB). Walking around the

VA hospital, you'll wear it somewhere on your body or belongings, either pinned to a hat, embroidered on clothing, tattooed on an arm, or printed on the lanyard to the keys of your motorized scooter, a common form of transportation on the grounds of the hospital.

If you don't come home with a chest full of medals on your lapel, people wonder what kind of warrior you were. Did you play it safe and cower, as Clay did, while your fellow warriors, like me, risked their lives? Did you take assignments where you never had to be in the thick of it and risk taking a bullet? In other words, were you the most emasculating thing you can accuse a Marine of being—a coward?

Some Marines go to extremes to earn the CAR. I witnessed Marines and soldiers who would move to the sound of an engagement and sometimes blindly shoot in that direction, the rationale being "If I fire my rifle, I'll qualify for the award." Of course, the indiscriminate fire could also hit civilians or cause other collateral damage.

One company had so little exposure to possible threats that two Marines conspired to fabricate a troops-in-contact incident. They were standing on guard. One moved ahead of the post and threw a grenade in the direction of his buddy. Then they both returned fire in the direction of the detonated grenade. If grenades were not a controlled item, these two Marines may have gotten away with falsifying an enemy contact and undeservedly earning the CAR.

Without the CAR on your lapel, or at least a CIB or CAB, many vets return home to irrelevance. Seeking another opportunity to earn that precious ribbon can be another reason why some vets wish to return to combat.

The glorification of war begins at the get-go. In the military, you learn about famous generals like Patton, MacArthur, and Eisenhower, among other war heroes. Some of the cadence songs sung during runs would include the names of Marines like Lewis "Chesty" Puller, who'd been awarded the Navy Cross—the second-highest medal for heroism—five times. These heroes are inescapable, because almost everything on every base is named after them. You set sail on ships named after battles.

In popular culture, there are few movies about a successful tour of duty without combat action. Instead, they tend to make movies about the heroes who went above and beyond the call of duty.

In boot camp, they read you Medal of Honor citations and tell you how most didn't come home to families greeting them with flags waving—instead, they came home in a casket. The mother or the widow accepted the award "on behalf of a grateful nation."

Given the chance, I knew I could get a prestigious combat medal. I was determined to do that. So were virtually all the other Marines I knew.

As I mentioned earlier, upon discovering they were not going to a unit deploying to combat, many young Marines at the school of infantry would purposely fail their last exams to get a different set of orders. Lieutenants I knew would try to shift from non-deploying units to deploying ones. I knew two senior officers whose sons had had their orders changed to allow them to deploy. Whether or not they had asked their fathers to get involved, or whether others thought that by helping their sons out, they'd curry favor with the generals—I don't know. Regardless, everyone who could call in favors to deploy seemed to be doing so.

Fourteen years after his retirement, one vet I worked with still felt like an imposter, because he was an operative in paramilitary operations for another government agency (OGA) rather than an infantry guy. The paramilitary OGAs lead and manage covert action programs and collect foreign intelligence vital to national security policymakers. Because they cannot share with others what their roles were while in the military, they carry no visible sign of being a hero. Without a medal on his lapel, this veteran felt that his contribution was not valid.

Many veterans who returned home felt the true heroes to be the ones who didn't return. They wish to redeploy to ensure that these heroes did not give their lives in vain—to ensure that their sacrifices still have meaning.

After Clay prevented me and my Marines from getting the CAR at first in 2003 for our combat experience in Iraq, I returned home without hero status. It was humiliating. That shame was likely a primary reason for me wishing to redeploy.

My career was not yet over, however, and I did finally receive the CAR. Others cannot return to service, either because of retirement or injury, and as a consequence, the chance for a combat metal is gone. Without validation for service, hope is lost, and many fall into a hole that is hard to dig out of. To self-medicate, they turn to drugs, alcohol, risk-taking behavior, and violence.

Some envy their buddies with lost limbs but a chest full of medals, and they think that they would feel better if they, too, had come home with a missing limb, paralysis, crippling back pain, or a disfigured face. Such visible injuries would have at least reduced the shame of not continuing to deploy, of no longer having a chance to contribute.

How much of your toe, foot, leg, or arm would you give up for a Congressional Medal of Honor or a Purple Heart? Maybe that sounds extreme. But it's still a trade considered within the veteran's mind, and something we often discuss in my groups. Would they trade another veteran's physical injuries to have had that veteran's combat exposure? I had one young veteran who had considered if he would trade places with a friend of his whose back was broken while supporting special forces in combat. Ultimately, this veteran determined that he'd welcome the life of painkillers and physical therapy for the opportunity to be a combat vet.

Even if they've received a medal, a paragraph on an award citation, or perhaps a folded-up flag, many vets who fought in Iraq or Afghanistan feel unappreciated for their service and sometimes imagine or assume that their buddies left behind resent them for discontinuing service.

Even if a warrior never gets a medal for service, their service nonetheless gave them something invaluable that could make it all worthwhile: a tribe. In the next chapter, we will explore this yearning for a tribe among service members.

CHAPTER 15

SEEKING TRIBE

Men do not fight for flag or country, for the Marine Corps or glory or any other abstraction. They fight for one another.

—William Manchester

The day I came home for good from the military, Amy and my kids greeted me at the airport with American flags and balloons that read "Welcome Home, Dad" flying from their arms.

Hugging Amy and feeling her body next to mine was a strange feeling. I was used to body armor and a rifle slung across my body. I tried to smile, but it was pasted on.

The minute I got a break from the hubbub, I called some of my Marine buddies. To Amy's annoyance, I arranged to see them at a local bar the next night.

"You just got home," she said. "Your kids want to spend time with you."

"I need to go."

"Isn't it good to be home?"

"Yeah," I said, not looking at her. *No!* my brain was screaming. *Fuck all this. I want to go back.*

"I get it," she said. "Go be with your buddies."

One important drive for vets wishing to return to service is the opportunity for brotherhood. When you're in the military, you feel like you're part of a big family. Each warrior faces the same dangers, the same problems, and the same goals. You relied on each other to stay safe. You tolerated the stress of being in danger in large part because of the bond you formed with your buddies—*all for one and one for all.* If a brother or sister you served with called and said, "Hey, dude, we need you here," you would go back to combat in a heartbeat.

At home, people don't know their neighbor, and texting or Instagram has replaced face-to-face interactions. "No man left behind" becomes "every man for himself."

In his book *Tribe*, astute journalist Sebastian Junger ponders why someone who went through this nightmare experience of war would find they actually miss it, and he attributes the reason to brotherhood. What you miss is not the blood and guts. You miss being in a tribe of small bands of people that depend on each other for survival. As part of that group, you develop deep, enduring social ties that protect, bind, and even heal.

When you come home from war zones, whether you see combat or not, you miss the brotherhood, the sense of sacrifice, and the mission that comes with war. The memories of your brothers and sisters tend to be the deepest and most long-lasting. I made it back, enabling me to keep those memories; some weren't as lucky. And sometimes, you are able to return because those who didn't saved your life.

Because of this very close connection to those with whom you've previously deployed, it's hard to leave them. You want to be there if your unit deploys again. Letting your buddies risk their lives again in combat without you would be too painful.

When I came home from Iraq, I didn't like to socialize with civilians who didn't have a military background. They weren't a part of my tribe. Though I had a large family—my wife and kids, my parents and in-laws, two living brothers, and two half-sisters—I longed for the camaraderie I had with my "band of brothers." Vets even refer to each other as brothers and sisters. Only with military friends do I feel a near immediate bond, and these are the guys I want to hang out with. Having enough veterans for friends seems to placate some of my urges to return to service.

Of course, now the bulk of my cohort are also retiring or ending active service. My ties to the active-duty ranks are dwindling, but my urge to return lingers. Working at the VA helps quell this desire to some extent. But I constantly have to watch my tendency to put veterans and patients ahead of my family, as I did when I was active.

Once, I met non-military friends at a bar, who I had known before I joined the military. "What was it like?" they wanted to know. I didn't know what to tell them. I worried disclosing too much about my combat experience would create misunderstandings and judgment.

After downing several beers, I loosened up and relayed the story of when I was the vehicle commander at the front of a small convoy, traveling with our transition team, and our vehicle was hit with an IED. After realizing what had happened, we thought we knew which vehicle had placed it.

"There were only nine Marines, myself included, and three vehicles," I told my friends in the bar. "We were operating at the absolute minimum possible. This meant that once we ran down the vehicle, I would be the only Marine who could get out of the vehicle to search the Iraqis."

Their mouths were open as they imagined this scenario.

"After doing the quick math in my head, I decided only the interpreter and I would advance on the vehicle. We did and came at them from the flank. At about twenty-five yards, I realized I would have to search them for suicide vests or weapons. No one would be there to cover me because my interpreter wasn't armed. I had a rifle and pistol, so I put the rifle selector on fire, handed it to my 'terp' and said, 'Cover me from an angle. It's on fire. You just pull the trigger.'"

"No shit!" said one guy.

"I closed the distance with my pistol out before I holstered to search the two men. I didn't find anything. Although there was a hole in the bottom of their truck where they could've placed the IED, and they claimed to be lost without having directions for where they were headed, I had nothing to detain them for and let them go."

"Man, you must have been shitting in your pants," said another guy.

Despite these guys listening intently with eyes wide open, I didn't feel they understood the breadth of what I had conveyed. How could I possibly express the shock, fear, anger, frustration, and shame I felt all at once during that incident? No civilian who hasn't stared death in the face could get an inkling of the hodgepodge of emotions, reaching a level ten, that I felt in that moment all at once. Only other vets who have been in combat could understand.

While I have a large family to support me in reintegrating back into civilian life, not all vets are so lucky. Many don't have families to return to; some grew up in an orphanage or foster care home, or they have a family too overwhelmed with their own problems to support their return. Disconnected from the peaceful world of civilian life populated by people who haven't experienced walking in the shadow of death, they lose the sense of belonging they felt in the military. Their desire to return to combat is actually a desire to return to a place where the people around you understand you.

Feeling profoundly alone and alienated, they plunge easily into depression and often display self-destructive behavior: alcohol, drugs, dangerous risk-taking, and even suicide. In the movie *Thank You for Your Service*, based on stories of soldiers coming home after deployment from Iraq, a soldier discovers his fiancée has emptied their apartment and moved out. She won't answer his phone calls. Finally, he confronts her in the bank where she works. She tells him it's over and asks him to leave. He puts a gun to his head and pulls the trigger.

Vets become alienated not just from family and friends but from civilian society as a whole. To them, civilians lack the honor, courage, and commitment of their brothers and sisters, something many have shared with me in therapy. Unable to connect with the non-military, they don't participate in civilian social activities. Furthermore, skills gained in military service may not translate well into civilian society. Finally, the teamwork they were used to may not exist.

This social alienation from society presents a huge problem for returning vets. War experiences don't disappear after you take off the uniform. Successfully readjusting to civilian life after war relies on combat veterans connecting with other combat vets after coming home. It seems like only another vet can validate your war experiences and help you not feel crazy for feeling the lingering effects of war. For this reason, many vets I've worked with prefer a veteran mental health worker, because they can only trust another veteran to understand them.

While the need for tribe and fame drives many of the vets I work with, so does the need for the excitement experienced in war. Let's now explore this dynamic.

CHAPTER 16

SEEKING HIGH

He remembered the firefights of his first deployment. "I loved it. Anytime I get shot at in a firefight, it's the sexiest feeling ever."

—*Thank You for Your Service*

War is intoxicating. Not just physically, but psychologically as well: the adrenaline produced by combat can be as addictive as heroin. This can make life between deployments and after discharge from service meaningless and boring, and veterans reenlist to get that high.

Officially, as we learned, "combat addiction" is a diagnosis coded under PTSD in the DSM-5 under which returning vets seek the high they experienced in combat. Seeking sensation, they get a high from staring death in the face, because that experience infuses them with dopamine, the reward neurotransmitter. Such sensation-seeking behavior is hard to curtail because of the excitement it generates. In the book *Generation Kill,* one nineteen-year-old corporal compares driving into an ambush to a *Grand Theft Auto* video game: "It was fucking cool."

When vets come home, they seek ways to chase that rush. They hunt, go to shooting ranges, bet on dog fights, and attend car races. They drive fast motorcycles, get into fist fights, engage in thrill-seeking sports like skydiving, and otherwise put themselves into dangerous situations over and over again. To recreate the thrill of combat in the workplace, many go into law enforcement, become security guards or firefighters, or work for the FBI or CIA.

One veteran, a former Vietnam pilot, owned a bar and restaurant. The crowd was young and violent. There were lots of fights, brawls, and thrown glasses. He was in his element and loved it. When he went home, he felt unsettled, angry, and miserable, and he looked forward to that other life where he could get back on the edge.[xiii]

Law enforcement especially is replete with combat veterans because it provides a similar experience to combat. They carry a gun, wear a uniform with medals or badges, have the comfort of structure and job security, and feel the toxic masculinity associated with the power of being in an enforcer role.

Reckless behavior is rampant among sensation seekers. One veteran used his status as a retired law enforcement officer to wander the streets of a medium-sized city late at night "just hoping someone would mess with me."

A highly decorated Vietnam veteran, who had been injured in a mortar blast, spent the first year stateside in a hospital before recuperating at his parent's rural home. He felt profoundly isolated without veteran comrades, and life seemed purposeless.

While drunk, he would drive recklessly on back roads after midnight to recreate the adrenaline charge of combat. He got a thrill from driving his car fast enough over the edge of a steep drop to become airborne, feeling the "bang" of hitting the pavement and the rush from seeing sparks burst in the rearview, all while undergoing the life-threatening struggle to regain control.

In a 2011 Northwest News radio segment produced by Austin Jenkins, Bridget Cantrell, a PTSD counselor in Washington who's worked with veterans for twenty years, says, "They have faced death so many times that there's no fear of dying, so they will push the envelope to the extreme to see how much they can deal with."[12]

As the radio piece describes, however, sometimes these thrill-seeking veterans push the envelope too far. It cites a 2011 U.S. Army report warning of an increasing propensity for soldiers to engage in high-risk behavior. The report asserts that this propensity led to 146 deaths of active-duty soldiers last year, including seventy-four drug overdoses.

"Criminal offenses among soldiers," Jenkins reports, "are also on the rise—nearly 75,000 last year Army-wide."

A veteran of a unit engaged in frequent combat in Afghanistan reported in a therapy session that, in the immediate aftermath of the Colorado movie theater massacre, he and a fellow soldier went to surrounding theaters nightly for a month. They were both armed with heavy-caliber semiautomatic handguns and staked out separate sides of the theater. They believed that the Colorado event would be imitated, and they wanted to be there to terminate the perpetrator.

When asked to explain his behavior, he said they were "going stir crazy" post-deployment, and so they emulated the search-and-destroy activity of combat. "I know a whole bunch of guys doing the same thing," he added.

The key to retraining for vets is to give their life meaning. Unfortunately, sensation-seeking warriors find purpose in dangerous thrills. Little else seems to provide them with enough oomph to replace the combat high.

CHAPTER 17

SEEKING PURPOSE

So the Marine Corps really did teach me to conquer fear, and then to go for higher causes, higher purposes.

—Robert Kiyosaki

In the military, you feel part of something bigger than yourself, something many are willing to die for.

Deployments, especially combat deployments, were where you knew how to use the skills you had developed. What you did was significant, meaningful work.

When you leave the military, finding work is challenging for many, and returning vets feel a lack of purpose and meaning in their lives. When serving, you develop skills to progress in the military, not necessarily outside of it.

Second jobs are discouraged while serving, and in some fields, they are not authorized. While you're on active duty, you can't easily build a second skill that you can use once you're out. The best you can do is parlay your military skill set into contract work with security firms. Unfortunately, the purpose you had while serving in a military branch often gets lost as a contractor. Being a mercenary doesn't scratch the itch quite the same way.

Many veterans find a similar form of purpose by joining the ranks of first responder professions. Law enforcement officers, firefighters, private security firms, etc., are all filled with veterans. Many of these professions seem to scratch that itch by allowing the veteran to continue to be a protector. Lt. Col. Dave Grossman used a metaphor of sheep, wolves, and sheepdogs. The sheep are the civilians, average people who have no interest in violence. The wolves are the predators who want to cause violence. And the sheepdogs are those who protect the sheep from the wolves. The challenge is that to the sheep, the sheepdog looks a lot like a wolf and can provoke similar fear, making integration back into the community a challenge. Numerous veterans I've worked with identify and even tattoo the word "sheepdog" onto their bodies. Sometimes I wear a shirt that says "sheepdog" on the front and "looks scary, protects sheep" on the back. Finding that protection as purpose outside of the military is difficult. And as recently I've discovered, many veterans may not be sheepdogs at all, because the truth is, we were trained to be wolves.

I was a wolf, trained to cause violence to the enemies of the United States. I thought I was a sheepdog and that being a protector was my purpose. I retired officially in 2014 and floundered for years in search of employment to replace the war fever of my combat experience. Eventually, I satisfied this need by becoming a clinical psychologist to help other vets adjust to civilian life. This gave my life purpose.

I was lucky. Other vets never find purpose in their lives except by redeploying.

PART V

RETRAINING

CHAPTER 18

BECOMING A PSYCHOLOGIST FOR VETS WITH WWS

I t was 2014. After having been with JSOC for three years, it was time to retire, and I still had not done any "traditional" combat rotations. Although I had raised my hand for every mission that came to our unit—four months in Afghanistan, four months in Yemen—most had fizzled out before deployment. Now it was time to leave the service, but my desire for combat remained frustratingly unsatisfied.

I pondered briefly what factors might be driving my addiction. Then I had to shelve this personal exploration. I was struggling more with the challenges of finding a new job and supporting my family. In my mind, the anxiety, anger, and confusion I felt upon retiring were owing to my unstable situation and not my frustrated urge to return to combat.

By 2016, I worked as a civilian with JSOC contractors and SOF operators in evaluating the candidates for special mission units. Most had distinguished themselves by serving with units in the Navy SEALS, Army Special Forces, or Army Rangers.

There was one candidate I identified as a narcissist. At first, I suspected it because of his smug attitude. Later, a mistake he had made during the training exercise confirmed my hunch. Not adhering to advice and rules, he got us stranded on an airfield with no way to escape. I had to call in for support to get him out of there. It was a complete mission failure. Another time, he was told not to drive his vehicle off the road but did so anyway, as if the rules didn't apply to him. The vehicle ended up stuck in the mud, and his entire mission was once more ruined.

Given his performance, I would not recommend him for the next level of training. However, because I was a journeyman not a psychologist at the time, how could I justify narcissism as disqualifying him from work in JSOC? I could not diagnose him with a personality disorder; I was just a former operator who knew something was off about the candidate and did some quick Google research on "narcissism."

Craig Bryson, an operational psychologist at JSOC, was there as a consultant in evaluating the mental health of the candidates. A medium-height, bespectacled, academic-looking guy with short, slightly graying dark hair, I knew him from my time on active duty, when we had a conversation about how my training was making me belligerent, often wanting to murder jerks who irked me. He concurred this was a normal military response and not psychopathological—assuming I didn't actually attack a guy with a murderous choke hold.

Craig supported me in my decision not to recommend this candidate. "Include the problematic symptoms in the write-up," he told me. In a letter to the commanders, Craig would state that while this guy was a great soldier, he was not what the unit was looking for at this time—not a failure, and not a good fit, either.

Craig and I discussed my desire to help vets like me who struggled after their service.

"Why don't you pursue a doctorate degree in clinical psychology? You can relate to these guys in a way that I never can. When you hang out with them, they know you've walked the walk, you've done the deployments. You're not just some doctor who sometimes wears a uniform."

The thought was appealing, but I had mixed feelings. It would involve six years of school, internship, and residency. I was thirty-eight with three kids and another one on the way, and I wouldn't finish my degree until I was forty-four, consequently not getting my license until

I was forty-six. Until then, how would I contribute to my family's income?

Still, becoming a psychologist and working with vets made sense. I could help them with their struggles in retirement and, at the same time, gain insight into my own struggles.

I discussed the idea with Amy. She thought I would make a great therapist.

"That's your true calling."

Craig wrote a letter of recommendation for me, and I applied to Argosy University's program in clinical psychology, the closest doctoral program to my home in California. In my second year, however, the university went bankrupt, and I had to switch to the Chicago School of Professional Psychology for my classes.

The schoolwork was interesting, and the path seemed correct for me. In the first year, I learned the technical terms for some of the struggles that I was having at home, like identity confusion. Academia helped to start me on a healing journey.

Although I was pursuing a clinical psychology doctorate, I was also required to take a social psychology class. This class introduced me to social field theory developed by Kurt Lewin, a Gestalt psychologist, in the 1940s. The theory posits that our individual traits and the environment interact to cause behavior. If every behavior is a result of the person and environment interacting, then the sum of a person's behaviors is their personality.

The idea resonated with my experience. To understand a veteran, you need to understand their experience in the military—the two are intertwined. The environmental factors associated with entry-level training in the military and subsequent deployments were enough to erode the normal human reactions to stress, for example.

While I've often described in this book how the military changes you to become a "bad ass motherfucker" (BAMF), let me offer a particularly telling example. The Marines used a "kill house" at the base in Twentynine Palms. During training, they rushed into the house after a squad patrol to find amputee actors dressed in fatigues and body armor, with fake blood everywhere. Seeing a woman holding mannequin legs as if they were her own alarmed me, even though I knew this was coming.

The Marines had to triage and treat these actors while the screaming and fake blood continued to pour out all over. Afterward, a quick debrief determined that the Marine Corps was now satisfied that these Marines wouldn't freak out when this happened for real. It actually worked. In 2007, after an IED strike, I had a Marine who had three limbs traumatically amputated. Desensitized from their training, his buddies were able to put tourniquets on all his limbs without hesitating. (Unfortunately, he didn't survive.)

Such desensitization, however, has a negative aftereffect in civilian life. My numbness toward horror, pain, and suffering had changed me from the patient person I had been before the military into an irritable father smashing my kid's toys whenever they annoyed me. The military had broken me, I thought. But with my newfound knowledge gained from my doctoral studies, I switched my mindset from "broken" to "needing training." I needed to learn more appropriate responses to the different magnitudes of stress in civilian life, just as I had taught Marines to do in preparation for the combat environment. I could do that.

The impact of environment in the military on forming personality came to light especially when working at the Children's Hospital of Orange County with at-risk adolescents in an intensive outpatient program (IOP). A first responder father tried to save his daughter from self-harm by approaching the situation as a problem to be solved, the mindset he had learned in the military. Drilled into us was the belief that if we focus on getting the solution, all will work out. Stuck in his rational mindset, he would ignore his daughter's feelings and tell her what to do. Misunderstood and invalidated, the girl would flee to her room, grab a razor, and cut her wrist. Such suicidal gestures landed her in the emergency room and then in the hospital's IOP.

To help treat the daughter's self-destructive tendencies, I used dialectical behavioral therapy (DBT). A long-term therapy commonly used with suicidal teens, DBT promotes accepting negative emotions, feeling them, and letting them go. I used distress tolerance skills, including techniques like self-soothing and distraction, to teach her how not to feel so emotionally overwhelmed. To help the father and daughter interact more effectively, I worked on their interpersonal skills.

The therapy with the parent-child dyad was a success, and it gave me insight into my own behavior. I realized that I, too, was not listening to my emotions—that I, too, treated every event as a problem to solve. I had been mission-focused for so long that any issue presented to me was merely a task to be completed, with no room or time to consider emotions. How someone feels when faced with an artillery strike matters not at all.

A real "aha" moment for me came after being handed an "emotion wheel" at the children's hospital. This was a circle with primary emotions, like sadness, joy, and guilt, in the center. Iterations of these primary emotions appeared as spokes so that more nuanced feelings were on the edge of the circle. I learned that anger is typically a secondary emotion, meaning it's the expression of some other emotion, like frustration or sadness.

I thought about how, having grown up playing contact sports, I had assumed the role of alpha male. Needing to be the tough guy, I was only allowed to be happy or angry. Crying was *verboten*. Only occasionally, like when I had a serious injury, did I shed a tear. And then it was "rub some dirt on it and get back out there."

These suppressed emotions instead manifested as anger in the military. Frustrated that someone else got picked over you? Punch a wall or strike another person during martial arts training. Feel guilty about having done something wrong? Blame someone else and yell at them. Anxious about an evaluation? Take it out on the troops who serve below you, like Marone and Clay did. In the military, everything was hardcore, vicious, and intense.

Gradually, I learned that I didn't have to bounce between presenting myself as "fine" and smashing my kid's toys.

At the IOP, I connected with the adolescents and their parents, and I made a difference in their lives. One patient even said he wished I was his father—high praise that made me proud. Implementing coping skills with the teens worked great for the most part, but something was still lacking. The therapies worked for solving the immediate problem and helped the teen toward being released from the hospital, but they didn't get to the core of the problem: the existential meaning, purpose, and value of human life. Seldom did I help the teens figure out what made their lives worth living. What was their purpose?

Then I had a breakthrough with one of the teens. He enjoyed calculus, so I wrote down Kurt Lewin's mathematical equation for behavior, expanding it to personality. Every single behavior is the result of some function that combines the variables of the environment and the personal. Behaviors could be plotted over time and described as a curve on a graph. The summation of all of my patient's behaviors would be the area under his curve from time "zero" to time "now." That sum was positive; therefore, his life was worth living. After many existential discussions about the meaning of life, we were able to work through his nihilism.

Reading more about existentialism, including Frankl's book about meaning in life, convinced me that finding purpose was the ultimate goal of life's road trip. If someone had a clear goal, they'd be willing to change their vehicle's tires, engine, and everything else in order to get there.

In applying it to the military, I felt there was no limit to the amount of suffering a warrior would endure to achieve the mission, a purpose in life that worked well in service. But I was no longer a warrior. And just focusing on getting my master's degree, doctorate, license, etc., wasn't enough to satisfy that desire for purpose in life. I needed more for my life to have meaning.

My direction was not yet clear. While incipient thoughts about WWS streamed through my mind, I had not yet thought about developing my ideas. Becoming a child psychologist was the right path, as I could save kids before they entrenched themselves in their psychopa-thologies by adulthood.

I applied for another program with the children's hospital for my fourth year, thinking that eventually this would lead to an internship position there. But I ended up working with children and adolescents with obsessive-compulsive disorder (OCD) at UCLA. The gold stan-dard treatment for OCD is exposure and response prevention (ERP). This requires putting the patient into a heightened level of stress and then helping them find other methods to reduce the strong emotions associated with their obsession instead of relying on their compulsions. For instance, if the child had to wash his hands ten times before eating a meal, we would force the child to forego this ritual, though that caused intense anxiety, and instead practice deep breathing. Though this treat-ment at first seems harsh, the method effectively reduces the compul-sions over time.

I believed that if I applied this same thought process to my own training and WWS, I could use the same principles to reduce with-drawal symptoms experienced after service. This fired me up, and I wrote a brief proposal for a qualitative dissertation to examine WWS. The proposal, however, was rejected because there was no literature on the matter. Instead, I started to write my dissertation on how effective IOPs were at reducing the distress of the adolescents.

Then COVID lockdowns happened. The children's hospital IOP for at-risk youth had to go virtual. My classes for my doctorate switched to all virtual. I set up my bedroom as a virtual classroom and used the dining room of my in-laws as my virtual clinic. I worked for UCLA the whole year without traveling once from my home to the campus at Westwood. It was surreal, and I loved it. I had found my niche. Thoughts about working with veterans drizzled away.

Around this time, some of my Marine peers were also retiring. When they called, we'd get into a discussion about regrets. Did I have any because I left early? Did I have any about not becoming a battalion commander or general?

I did at first, I told them. I felt like I wasted six months of my time between 2013 and 2014 going to command and staff college, a requirement for promotion to lieutenant colonel. Since I was retiring, I thought that school time had been wasted. I could've been deploying to combat with that time.

Then it occurred to me that deployments would have been meaningless. The war continued anyway, and I would only have been exposed to more harm. Instead of regretting that my staff college time had prevented me from more combat, I should have seen it as having prevented me from making something of my life outside of the military.

That started me thinking. Why was returning to combat more important than learning more about myself—more important than spending time as my kids' soccer coach and setting date nights with Amy to strengthen our marriage? These family-related activities were more meaningful than staring death in the face and sweating bullets in a godforsaken desert or the mountains of Afghanistan.

WWS as a possible mental health syndrome was taking shape. While I had not found anything in the literature regarding combat addiction, I was aware of the "operator syndrome" in the military, part of which described the poor transition to civilian life and lack of purpose. This theory posites that some of the challenges faced during the transition from military to civilian life are because life outside of the military is more mundane and lacks the excitement of life-and-death situations. WWS was, in a sense, a mental health component of operator syndrome, although it didn't seem to require the deployments or special operations MOS.

I discussed the dynamics of WWS more and more with other veteran psychologists, who agreed the concept might be worth exploring.

Meanwhile, time was getting short for applying for an internship. I applied to multiple locations. With a near-perfect GPA, good letters of recommendation, and a wealth of experience from military service—not to mention working with adults and adolescents—I had a good pedigree.

The Navy had an internship program that would satisfy my doctoral degree requirements for a clinical internship. I sent in my application and tried to enter the Navy as a Medical Service Corps officer. But I ran into red tape. It is the military, after all.

In the end, the military entrance processing station (MEPS) looked at my medical history and told me that my disability rating prevented my return to service. While I had no traumatic injuries, I had knee surgery from jumping out of planes. Hard landings had led to damage between the bones. Also, because of an explosion that had gone off two feet from my head, I can't hear the five-hundred-hertz frequency in my right ear at all. Turns out that a stick of dynamite going off that close to my right ear permanently damaged its cochlear hair cells.

I applied to the local VA hospitals for an internship, as well as all of the children's hospitals in the area. I sent out over thirteen applications. My top choice was to go back to the children's hospital where I'd worked in the IOP. I only got interview requests from that hospital and one other children's facility. All of my internship eggs were now with these two sites. The national Match Day for assigning residencies came and went, and I hadn't matched anywhere. This was the fourth time I'd been rejected as a psychology applicant, counting the two times I'd been rejected for practicum sites as a student. It was quite a blow for someone who'd excelled at everything before and for whom being "last pick" was unknown.

Ultimately, I got an invite from San Bernardino County's Department of Behavioral Health (DBH). I was interested in community mental health, and they had both a transitional youth program and an outpatient clinic treating children and adults.

I matched in the second round with the DBH and ended up working with adults and children with serious mental illness. I asked for the hardest children's cases that the clinic had, specifically working with young boys. Since the clinic had few male therapists, I was a good match.

I took the cases that others didn't want or struggled to manage. Patience learned through combat helped me handle the strong emotions of those young boys, and as a result, many of them showed progress. It was a difficult clinic to leave, but I had to consider where I would do my postdoctoral fellowship. Unfortunately, the DBH didn't offer one.

Still thinking I would be a child psychologist, I thought I would apply to the children's hospital again for a fellowship. Having improved my pediatric expertise in the last year, including diagnostic testing with some adolescents, I thought I had a fighting chance. As the children's hospital was close to my home, I didn't consider other postdoctoral fellowships over this potential opportunity.

Again, I didn't match, and at this point, I was frustrated. Where could I apply for a fellowship? One of my cohorts, a fellow Marine veteran, had matched his postdoctoral fellowship at the VA at Long Beach. As it turned out, they had another opening.

I applied and was accepted. I would be working with groups and conducting individual therapy. It was also an opportunity for me to further develop my ideas about WWS.

CHAPTER 19

DIAGNOSIS AND TREATMENT AT THE VA

Just because you can doesn't mean you should.

—Various

Vets enter the mental health system in different ways. Some come voluntarily to file a disability claim for anxiety, depression, or PTSD. Others enter the system because they were sent to the hospital for a suicide attempt or even homicide.

Following the first aid mantra that directs health professionals to "stop the bleeding, start the breathing, treat the wound, treat for shock," the VA has the capacity to do the first three steps:

First Step: Stop the Bleeding

In first aid, nothing else matters if you can't stop the bleeding because the body has a limited amount of blood. In mental health, this emergency stage can be represented by the inpatient and emergency services realm. The veteran is slipping away, and something has to be done to prevent the maladaptive coping mechanisms that lead him to suicide, overdose, etc. While inpatients, the veterans will be held by the VA until they are able to "start the breathing."

Second Step: Start the Breathing

Once the bleeding is stopped, we have to make sure that the blood is being oxygenated, or it's not helping the organs. This is the "start the breathing" phase, which, in my VA analogy, refers to the transition from inpatient or intense outpatient to outpatient services. It requires getting buy-in from the veteran that outpatient psychotherapy can be helpful and gaining their commitment to attend.

Unfortunately, this step can take some time; the veterans I would triage had their consults put in three to four months prior. In the biographical movie *Thank You for Your Service,* three soldiers return home from Iraq with severe PTSD. One of the soldiers, Solo, suffers severe PTSD and memory loss. He waits hours at the VA before he gets to speak to someone about making an appointment for treatment. When it's his turn, he's told there's a waiting list and it could take months before he sees a counselor. Fortunately, a counselor nearby hears his complaints about how badly he needs treatment and pulls him into her office.

After any veteran's consult, it could take a few weeks to months before they were linked back to a provider and could begin their "treatment of the wound"—that is, if the veteran hadn't dropped out already.

In my own practice, one triage veteran patient had his consult put in the year before, but because of his own scheduling conflicts, he continued to miss the appointment and wound up being rescheduled for four months later, resulting in a new appointment thirteen months from the initial consult to our meeting. When we spoke, he scoffed at me as a VA representative and stated that he didn't need our help anymore. Over the course of the year, he claimed he had "figured it out."

Third Step: Treat the Wound

Treating the wound should include stabilizing an injury, covering it to prevent infection, etc. Basic life support has been managed at this point, allowing for some work on the immediate injury. In my analogy, this step is represented by the short-term outpatient therapy.

"Treating the wound" also requires that the veterans actively do their part to effect change. That means attending sessions and completing the homework I assigned. Homework might consist of trying to use a new coping skill instead of their previous maladaptive skill. For example, a patient might sit with their palms facing up on their thighs, attempting to use a "willing hands" posture to increase their willingness to try something new. Other homework assignments include things like reviewing interpersonal effectiveness skills such as assertive communication. I sometimes offer handouts that help the veteran review the way a skill works, prompting him to think about how he felt before practicing the skill and how he felt afterward.

When the veteran begins treatment at the VA, the majority of that treatment is short-term and strictly goal-oriented. The VA wants to offer enough services to keep a patient's head above the water's surface. At the same time, some of the programs do have longer terms, and after one four-month round of psychotherapy, the veteran can ask to extend.

Treatment Modalities

Veterans present with different issues. Many will be diagnosed with PTSD, while others will fall under less extreme diagnoses such as chronic adjustment disorder.

PTSD Diagnosis

A veteran is likely to be diagnosed with PTSD if they present with hypervigilance, anger management issues, communication struggles, identity confusion, insomnia, thoughts about past service, concerns about moral injuries, and other related symptoms.

The challenge in making the diagnosis is determining if the veteran meets the DSM-5's criterion A: exposure to trauma. From the hundreds of patient files I've read, an objective metric for qualifying the trauma a veteran faces doesn't seem to exist.

Apparently, simply being in combat satisfies the requirements: the DSM-5 allows for "exposure to war as a combatant or civilian" as inclusive. That would mean I could meet the criterion A for having PTSD, though my combat tour from 2003 to 2004 in the Green Zone involved eating more ice cream than seeing action, with only one rocket attack yielding zero casualties over the course of four months.

Nevertheless, exposure to combat is considered for this diagnosis because this exposure can be traumatic, and because the warrior lives in a constant state of fear in this environment. When I was stationed in the Green Zone, living in a luxurious palace, I nonetheless lived in fear for my life, given that there was fighting nearby in Sadr City and IEDs being detonated along "Route Irish," the street going from the Al-Rasheed Palace to the Baghdad International Airport. Apparently, this fear was significant enough to qualify me for a PTSD diagnosis.

Even if not exposed to combat, military training alone can be traumatic for some. At the higher levels of survival, evasion, resistance and escape (SERE) training, the goal is to convince the service member that their life is in danger. From isolation and sensory, sleep, and food deprivations to physical assaults, including waterboarding, the student feels as if they are being held indefinitely in incarceration. Surely these veterans had endured "threatened or actual physical assault [including] torture [and] incarceration as a prisoner of war," as the DSM-5 specifies. Any one of these seems to be sufficient for criterion A of PTSD. Going to combat and being shot at is not required.

Among the hundreds of veterans' charts that I reviewed while on fellowship, veterans met criterion A from events including combat action where their peer next to them was killed, non-specific attacks on bases where the veterans were, motor vehicle accidents while on active duty, and military sexual trauma (MST). Some didn't mention a specific event.

Treatment

Veterans who are diagnosed with PTSD are triaged to a clinic or to a provider who will use the industry-standard evidence-based practices (EBPs) for treatment. These will likely include the following:

- **Cognitive Processing Therapy (CPT).** A twelve-session treatment that teaches the vet how to evaluate and change upsetting thoughts about the trauma. By changing your thoughts, you can change how you feel.

- **Prolonged Exposure (PE).** Veterans are taught to gradually approach and address traumatic memories, feelings, and situations. By confronting these challenges directly, veterans may see PTSD symptoms begin to decrease. PE typically lasts eight to fifteen sessions.

- **Cognitive-Behavioral Conjoint Therapy (CBCT).** This therapy is designed to help couples understand the effect of PTSD on relationships and improve interpersonal communication.

- **Eye Movement Desensitization and Reprocessing (EMDR).** EMDR involves calling the trauma to mind while paying attention to a back-and-forth movement or sound (like a finger, light, or aural tone moving from side to side). It helps vets process and make sense of their trauma.

These therapies for PTSD are designed to manage immediate concerns and provide some relief to the most distressing symptoms.

Short-term goals might include reducing the negative impact that the traumatic event has on a spousal relationship, developing better strategies (meditation, visualization, deep breathing, etc.) to cope with symptoms, and using anger management as an alternative coping strategy to avoid the negative consequences of anger.

Shortcomings of Therapy for PTSD

While the VA has many programs to help vets suffering from PTSD, there remain many shortcomings of therapy alone.

Not Designed for Veteran Trauma

While EBPs have proved beneficial for many vets, they seem to be designed largely for people who have experienced a single traumatic event like an IED strike, one firefight, or one death of a comrade, though this is not backed by research. Let's say your car hit debris and flipped over in the middle of the desert, and since then, you've avoided driving. PE, CPT, or journaling about the experience has been shown to reduce the hyperactivation of the amygdala, our fear guard in the emotional midbrain, and with the use of these therapies, you will eventually get back behind the wheel and drive again.

In my practice, however, I've found EBPs less useful to veterans who have experienced multiple, repeated traumas, common to those who have deployed multiple times. Also, these therapies do not address deeper WWS issues like identity confusion, lack of meaning, loss of tribe, and loneliness. For this reason, veterans with combat addiction or combat envy might find EBPs unhelpful and drop out.

In the article "Combat Addiction: Revisited and Reaffirmed," the authors note another reason for traditional therapies like PE to fail: trauma high. As sensation seekers, combat veterans often thrive in high-risk, adrenaline-pumping environments. High traumatic exposure reinforces high sensation-seeking behavior.

Another 2016 article notes that between one-third and one-half of veterans receiving exposure therapy (ET) do not demonstrate clinically meaningful symptom improvement. And even when patients do improve, symptoms often remain high.[13]

Anger Management

In group therapy and short-term therapy, little effort is often made to get to the root cause of the anger. This presents a problem in the success of the treatment because for most veterans, myself included, anger does not emerge from a slow simmer but rather a rapid explosion. In those situations, most of the coping strategies will not work because the person is in survival mode and will spontaneously fight for survival by yelling, punching the wall, or throwing a vase against the wall, among other responses.

Stigma Concerns

Veterans who receive a PTSD diagnosis sometimes refute it because the military will deny anyone reentry with this diagnosis. This means that those who want to continue to work for the government in the military, law enforcement, etc., may be denied the opportunity to serve. They may even be denied their right to keep and bear arms. The diagnosis could impact their chances of getting hired in other employment settings, as well. Who wants to work with a crazy Marine who might go "postal" and become extremely angry, to the point of rage and violence?

In order to avoid the PTSD stigma, veterans who need mental health services will aim for a different diagnosis, like generalized anxiety disorder or chronic adjustment disorder (CAD). To guide other veterans in avoiding a PTSD diagnosis because of this "mark on your record book," veterans have discussed these concerns on popular podcasts, including *The Joe Rogan Experience*.

WWS veterans often refuse to seek assistance, using maladaptive coping skills in place of EBPs, regardless of the divorces, life challenges, and job losses that may result. When I consult with reluctant veterans, I highlight that therapy is more a matter of retraining into civilian life rather than a sign that they are not fit for that life.

Chronic Adjustment Disorder

CAD is a stress-related mental health condition that affects the ability to form or maintain relationships or complete the tasks of everyday life, such as personal care, working, going to school, or providing for a family.

Many veterans develop CAD in response to stressful events or life changes. During combat, or even in a deployed non-combat role, military members are subject to stressful living conditions and may be under constant or perceived threat.

To diagnose CAD for military veterans, a licensed psychiatrist or other licensed mental health professional will note the veteran's symptoms and ask questions about incidents or circumstances that could have triggered the condition.

Some emotional and behavioral symptoms of CAD include the following:

- Impulsive or reckless behavior
- Agitation and anxiety
- Withdrawing from others and feeling isolated
- Crying or feelings of hopelessness
- Difficulty concentrating
- Changes in eating habits; rapid weight gain or loss
- Increased abuse of drugs or alcohol
- Suicide ideation or attempts at suicide

As can be seen from the symptoms, CAD captures many of the symptoms of WWS, with the exclusion of combat addiction or combat envy.

At the VA, I have treated veterans with varying diagnoses in both individual and group therapy. Let's now explore my experiences.

CHAPTER 20

TREATING VETERANS AT THE VA

If you choose not to decide, you still have made a choice

—"Freewill," song by Rush

A t the VA, my ideal goal was to treat patients who fit the WWS
profile.

To find veterans with the WWS symptoms cluster, I took referrals from triage appointments at the VA. These referrals came mostly
from veterans who were already seeing a psychiatric provider. Occasionally I got referrals from the urgent mental health clinic or from the
primary care mental health integration clinic.

Therapeutic Challenges

Because I was working in the behavioral health interdisciplinary program, most of the veterans referred to me recognized that they needed
mental health therapy. Having already "raised their hand" once, they
were motivated to get well.

Others, though, presented challenges, from a lack of awareness that
they needed help to feeling that mental health intervention was stigma-
tizing to previous bad experiences in other groups.

Nothing Wrong

Patients who denied anything was wrong with them were at the top of my list for inclusion in my group. They were often the ones with combat addiction or at least a propensity toward developing it. Situated in the pre-contemplative stage of WWS, they lacked insight and denied having problems. They might down a bottle of vodka a day to function but refuse to acknowledge that they have a drinking problem. They might say, "Drinking relaxes me," or "I hold my liquor well." This denial led them to happily sign away their identity as a well-adjusted civilian for the sake of satisfying their combat fix, continuing to seek deployments and look for guaranteed contracts, special forces opportunities, recon assignments, etc.

Despite having a marriage on the rocks, going on drunken binges, getting into bar fights, and snorting away their money, they were prevented by pride from seeing someone. Instead of directly seeking help, they might enter the system for something related to primary care, only to get referred to the mental health division.

Discontent with Groups

While some in my group were just starting therapy, others had been through the rounds and found that other groups came up short. For this reason, they had quit them.

One dropped from a combat PTSD group because it was led by a civilian who "had no idea what he was talking about. He just followed protocols blindly. No way will I talk to him. He just doesn't get it." Civilian therapists not understanding the veteran's experience was a common theme, particularly in some of the cognitive behavioral groups. One of them explained, "I was sick of hearing 'Can you please reframe that statement into a more positive vein?'"

Many did not share their desires to redeploy in a group led by a civilian therapist, feeling they would be misunderstood. If pressed, some couched their responses by saying they wouldn't reenlist but would serve if recalled to active duty.

"I get it," I said. "My buddies are all military. No one else can understand what I've gone through."

If they got "stuck" with a female civilian therapist, an unwillingness to participate was 100 percent. One said, "Nothing felt right with her," referring to a female therapist I knew who was perfectly empathetic and competent. His confirmation bias that "there's no way this chick will get it" led him to quit after one session.

Another said he was sick of hearing "sickening" tales of woe. These guys needed to just "bite the bullet." He would feel worse after a session than before.

Another quit because he felt like an imposter. "Most in my PTSD group had lost a limb, or a buddy. All I did was get a bullet hole through my helmet."

"Yeah," said a soldier who had lost an arm, "I quit because of fakes like you just looking to get compensation for a false PTSD diagnosis."

At this, the "imposter" stood up to confront the soldier, and I had to jump in to lower the heat.

"Whoa," I told them. "Let's keep cool."

I used this as an opportunity to get into a discussion about the inability to control anger because anger had been used constantly as a survival mechanism prior to civilian life.

I also added that I, too, had felt like an imposter because I hadn't been injured or lost a close buddy. "For the record," I told them, "I got a bullet through the notebook I was holding, and shrapnel scraped my helmet. That was why I kept deploying. I felt I hadn't really experienced the nitty-gritty of war. I was a fake warrior."

"Ooh-rah," said the one-armed soldier.

I sighed with relief. My sharing worked, and things quieted down.

Another reason for quitting therapy was a lack of generational identification. A seventy-year-old vet shared that he had nothing to contribute to the younger generation's experiences, while a twenty-year-old commented that he hadn't connected with the Vietnam era because their version of combat had been so different. What's more, the younger veteran didn't feel validated when the Vietnam veteran asked him, "How was it over in the sandbox?" That infantilizing expression seemed to trivialize his experience compared to the older veteran's time in Vietnam.

Long Wait Time

A sailor in his fifties felt that there should have been a better transition program to *de*program all the training he got from the Navy. Hearing him say this was music to my ears because it was straight out of the WWS playbook.

"They brought in some retired senior NCOs to teach us work skills," he told me. "That motherfucker never did any real work. He got out as an E-7 [sergeant first class] and went right to work teaching people like me about 'the real world.' He don't know shit about shit. That asshole never got a job working for a real company. If he couldn't do it for himself, how could he do it for me?"

The sailor had already done six months of therapy a year ago with the VA and was waiting to get reconnected for another round with a VA clinician. As the triage provider, I'd met with him for one session to either direct him to a group or make the referral for individual therapy. At the end of his previous sixteen sessions with an individual provider at the VA, he knew that he still needed help.

Because it had been more than six months since his last appointment with a mental health clinician, he had to go back through the triage process and get assessed for therapy again. Getting linked back to a provider would end up taking over a year for him. Over the course of that year, some of the progress he'd made had worn off, and he wound up back to where he had started. The "wound" wasn't treated, the bleeding started, and the breathing stopped.

Failure to Stick With It

Many WWS veterans I met coped with managing anger and volatile emotions by turning to the bottle, and they were encouraged to go on the wagon for at least two sessions to see how they coped with sobriety. Even though they felt some improvement, some believed they only had to "get through a rough patch" for two weeks, and then the rough patch would be over. They would discontinue therapy after that time and go back to those warrior-identifying behaviors that brought them to therapy in the first place.

Six months or a year later, they would return to therapy, realizing that things hadn't gotten better. I have had several of these cases in just the last year.

Preference for Meds or Experimental Procedures

Some vets found quicker solutions with medications rather than traditional therapy. I had a few of these cases while I was completing my fellowship. In addition, patients used different experimental procedures to help manage their depression or PTSD.

Ketamine Clinic

The VA has been working alongside the greater mental health community to study the use of ketamine for helping manage depression. Ketamine is a psychedelic drug (known as a "dissociative") that is undergoing a resurgence in popularity. Originally derived from PCP, or "angel dust," ketamine has been used in hospitals and veterinary clinics as an anesthetic or tranquilizer for decades, and it has been cited as a misused recreational drug known as "special K."

A prescription version of ketamine called esketamine was approved in 2019 by the FDA for treatment-resistant depression (TRD). With esketamine, relief from the crushing weight of depression often happens rapidly, within about forty minutes. According to the FDA guidelines, however, it is only to be used "under the supervision of a health care provider in a certified doctor's office or clinic." That means medical professionals need to supervise its use and then monitor the patient after they have taken a dose, checking vital signs and the user's clinical status.

One of my patients, who had cycled in and out of individual therapy with different psychologists before landing in my office, elected to try the ketamine clinic. After treatment, his suicidal thoughts dissipated, and he discontinued individual therapy with me, feeling he had learned the coping skills he needed now that his depression wasn't "as strong."

Stellate Ganglion Block (SGB)

Even before I retired from the military, my operational psychologist, Craig Bryson, was recommending SGB for active-duty military members and veterans with treatment-resistant PTSD. He found that all his veteran patients experienced significant relief with it.

The stellate ganglion is a collection of sympathetic nerves in your neck, found on either side of your voice box. You can get an injection called a stellate ganglion block, or sympathetic nerve block, to ease pain in your neck, head, upper chest, and upper arms. It can also help with circulation and blood supply to your arm.

One of my patients was receiving the injections. While it helped elevate his low moods, its effects quickly wore off. But in the meantime, the reduced amygdala activation gave him a break from depression and time to work on distress tolerance, emotion regulation, and interpersonal effectiveness. I told him that the SGB had brought him "inside the wire," where he could work on retraining his immediate action skills. Our goal would be to use that inside-the-wire time to make sure he was ready before going back outside the wire. And this meant he had to do his homework, responding to future distress by relying on adaptive coping skills developed during his time inside the wire. The work is tough for these veterans.

In the case of my patient, he was so pleased with the SGB results that he dropped out of therapy and instead sought an increase in SGB frequency treatments.

Repetitive Transcranial Magnetic Stimulation (rTMS)

TMS is a procedure that uses magnetic fields to stimulate nerve cells in the brain to improve symptoms of major depression. It's called a "noninvasive" procedure because it's done without using surgery or cutting the skin. Approved by the FDA, TMS usually is used only when other depression treatments haven't been effective.

During an rTMS session for depression, an electromagnetic coil is placed against your scalp. This coil delivers magnetic pulses that stimulate nerve cells in the region of your brain involved in mood control and depression. It's thought to activate regions of the brain that have experienced decreased activity during depression.

One patient I was treating who had had suicidal thoughts experienced significant relief while doing rTMS treatments. During the course of the rTMS, he continued working with me on managing his depression and anxiety.

Near the end of the treatment, he started to miss his rTMS appointments and then his appointments with me, as well. After missing two appointments in a row, he ended up in the emergency department with significant suicidal and homicidal thoughts. He was brought to the inpatient unit, and he stayed there until the providers determined that he would be safe to return to the community.

As with other patients I've treated with multiple hospitalizations, sometimes the goal wasn't to get to the root cause, or "treat for shock." Instead, success meant that the patient lived to see another day.

Individual Therapy

While working one-on-one, I had more veterans experience positive outcomes, manifested as significant reductions in their anxiety, depression, and anger, as long as they didn't terminate their therapy because they didn't immediately experience any noticeable relief.

One success story was "Joe," an Army veteran whose initial career was similar to mine, with deployments in Anbar Province where there were many casualties, including in his own unit. His friends had been killed by an IED, and in the aftermath, he suffered survivor's guilt. He felt that he should have been the NCO in charge of that patrol and not the corporal who had switched places with him during the time the IED had detonated. This guilt made him moody and frequently angry. He would explode at his wife over nothing but refuse to share with her what was troubling him.

Following active-duty service, he worked as a military contractor, returning to Iraq and providing security for Department of State officials. Working as a contractor put him in harm's way, satisfying that combat addiction. Still, for him, something seemed to be missing. Though the money was good, it wasn't enough to give him the purpose he had had serving the military.

Hearing him, I thought back to a sergeant who had worked for me in FAST. He had been with Blackwater, a private military contractor, before deciding he'd rather take a third of the pay in order to be a Marine again. He couldn't justify putting his life on the line for $150,000 a year, but he could while wearing a uniform for just $50,000 a year.

Joe did not return to active duty, although he considered it. His wife gave birth to a baby boy, and though the money from contracting would provide for his family, he felt obligated to stay at home.

He tried to go back to school to study business, and he worked different jobs, including one with an advertising company. Nothing gave him purpose or satisfaction. Ultimately, he was referred to me after the Department of Family Services was called to his house following an altercation with his in-laws. Because of his frequent outbursts, his in-laws had wanted his wife and baby to come live with them. When his wife attempted to walk out the door, he grabbed the baby from her arms. "If you don't butt out," he told his in-laws, "I'll take your daughter and my son to Mexico and disappear. You will never see your grandkid again."

Joe started the first few sessions with arms crossed, more focused on getting "credit" for having attended therapy than actually intending to benefit from it. He would ask for letters from me that he could bring to the Department of Children and Family Services (DCFS), proving his attendance.

We'd speak about the concepts in the WWS group, talk through how the military was a destructive relationship, and confront his warrior addiction.

After only a few sessions, we covered all of the WWS material and turned to more traditional EBPs to figure out how to reduce his distress, improve his emotional regulation, and work on his interpersonal effectiveness.

Using cognitive behavioral therapy (CBT), Joe learned to reframe his thoughts into a more positive arrangement. For instance, when he said, "I'll never find work that gives me any satisfaction," I encouraged him to reframe that. He did and would say instead, "There's work out there I can enjoy. I just need to find it."

I encouraged him to do extreme sports to get that combat buzz, and he took up bungee jumping. When he felt the rage about to explode, he would pinch his wrist, say "stop," and take in five full breaths. Within a few months, he was successfully using these and other adaptive coping skills. He tried teaching them to other veterans and thought about pursuing his own degree in mental health. His PTSD, anxiety, and depression self-report measures had dropped to nearly zero.

Failed Cases

During my fellowship year at the VA, I encountered three cases where the vets struggled to continue with me in individual and group therapies. In two of these cases, the vets terminated therapy before retraining could occur because they falsely felt that they had solved their problems. A third case may have committed suicide.

Therapy Not Needed

One of the patients who discontinued therapy was "James," a retired Marine who said he didn't need to "untrain" his military experience.

For him, working as a security guard provided the warrior fix. He wore a uniform, carried a weapon, and commanded respect.

Additionally, with his security role, he found a renewed purpose. He had a reason to work out again and to structure his life to be prepared for his security job.

At his job, he could be vigilant and stoic and respond to threats with violence. Because these were now acceptable parameters in his life, he was no longer interested in quelling them, even though at home he would end up smashing chairs, throwing vases, whipping his children for minor offenses, and, on several occasions, slamming his wife against the wall.

Denying his present difficulties, he missed sessions and struggled to work on his goals to control his rage. Believing that reconnecting to work would provide the warrior fix, he terminated therapy.

The second patient who terminated therapy was "Bill," an Army veteran who had been on two combat deployments. After those deployments, he decided not to reenlist for more combat tours, because he suddenly realized he could die.

In therapy, he made some strides. For instance, he recognized how he was still treating his family as if they were his troops. He would bark orders to them and insist on full compliance without negotiation. He was also aware of his regimented military behavior. We had a laugh about how he would lay out all of his equipment on his bed before he would go on a hike for fun.

"You gave yourself a junk on the bunk inspection?" In the military, a "junk on the bunk" inspection involved laying out all equipment required to bring to the field, after which the NCOs would go through to make sure the troopers had every item and then watch as the troopers packed them into their rucksacks.

"Yeah," he said, "I guess I did."

Fearful of being unprepared for catastrophic events, Bill kept trauma kits in his car. Even on fun hikes, he would have countless first aid items and extra rations. On one recent hike up a steep mountain in Montana, everyone seemed well-prepared, and he could enjoy himself. On another hike with a different group, some of the participants had "failed" to take even the most basic supplies. One "failed" to bring insect repellent and had to borrow it from another hiker. Another "failed" to bring extra batteries for a headlamp, which ended up dying. Another "failed" to bring an extra shirt, and when he got soaked from an unexpected rainstorm, he had to endure being wet for the rest of the hike. The chaos made Bill feel ill, and he barely had the energy to finish the hike.

We discussed his need to maintain a militaristic lifestyle, ready for anything, because it had paid off when he had been ambushed in combat. From my perspective as his therapist, getting him to relax more was a possible goal to work on.

But after that discussion, he never came back. Perhaps that level of insight was enough for him. I never found out.

Suicide Case

My third failed case was "Clyde," a forty-three-year-old Navy man who served in Iraq and was intent on committing suicide.

"Nothing you could do say or do would change my mind," he informed me.

Shrapnel in his leg had given him a limp and left him unable to redeploy. Without being able to return to service, he had nothing to live for except to take care of his elderly father.

He had made a pact with his father that he'd stick with him and try to shield him from his mother, who, according to him, was an "insufferable" and "manipulative" woman. She controlled the family's finances and would lock Clyde out of their accounts. She would also hide his car keys and make incessant demands on his time. He decided that once he had fulfilled his pact, he would choose between a few plans he had outlined to take his own life.

"What are these?" I asked.

"My first preference is with a gun. But that will leave blood. Hanging is second. I would do it from a hook in the basement that I use to attach workout equipment."

"Sounds like there's no way I could talk you out of your plan."

"No," he replied.

This plan was long in the making. Several years ago, he sent a letter to his civilian therapist about his plan, highlighting the things he thought were wrong in the world.

I completed the required suicide assessment scales and the suicide risk evaluation. Part of the evaluation was to come up with a prevention plan. My only recourse was to provide him with possibilities for finding a life worth living. If he couldn't find one strong enough to overcome his plans, then he was right: I couldn't crack his willfulness.

Having lost purpose after being unable to redeploy, he'd rediscovered it in protecting his father. When that purpose was gone, he had no reason to find another.

Clyde had no immediate intent to kill himself. His father was still healthy enough that he thought he'd make it several more years. Because he wasn't an immediate threat to himself or others, all I could do was discuss with him why he would have nothing to live for after his father was gone.

But after several conversations exploring other reasons to live beyond his father, nothing seemed to stick. Working on coping skills and using other EBPs wasn't helpful, either.

Clyde believed that his life had no purpose after his father died and that he therefore had no reason to continue. He had served his country, gotten married to a woman who—though he had never loved her—was devoted to him, and had two successful grown sons, one finishing an MBA and the other working with him in his restaurant serving Texas-style food. He had achieved everything that he wanted in his life and couldn't think of a reason to "stick around" afterward.

We discussed his belief that his life was meaningless, his extreme pessimism, and nihilism, which is a radical skepticism that condemns existence. He noted that his first civilian therapist had mentioned the word "nihilism" as well.

Because Clyde was a deep-thinking individual, I felt a philosophical discussion about existentialism and nihilism might stir him.

"Are you willing to do homework?" I asked him.

"No," he replied.

"What if it's just a few YouTube videos on absurdism?"

Head bowed, he was momentarily quiet. Then he glanced up. "I'll check them out," he muttered, half under his breath.

When he returned, we discussed how, from the absurdist perspective, the universe is irrational and meaningless, while the search for order brings the individual into conflict with that universe. Consequently, if he wanted to defy the universe, then the appropriate action would be to continue living.

I described Albert Camus's take on suicide from the essay "The Myth of Sisyphus." "There is only one really serious philosophical problem," Camus asserts, "and that is suicide. Deciding whether or not life is worth living is to answer the fundamental question in philosophy. All other questions follow from that."[14]

The answer, Camus states, is to live without escape and with integrity, in "revolt" and defiance, maintaining the tension intrinsic to human life.

Since the most obvious absurdity is death, Camus urges the despairing person to "die unreconciled and not of one's own free will." In other words, he recommends not a life without consolation but one characterized by lucidity, including an acute consciousness of and rebellion against its mortality and its limits. As another veteran I worked with put it, Camus is arguing that the veteran give a big middle finger to the world, essentially telling it, "I'm going to live out of spite."

These existential discussions were paired with trying to help him find meaning beyond his father—perhaps with his children or through adopting a pet. But nothing engaged him or deterred him from his plan.

In one instance of emotional vulnerability, he broke down when talking about adopting a dog.

"I've had my eye on a Dobermann Pinscher breeder," he said.

"Does it have to be a Dobermann?"

"Yes."

"What makes you say that?"

"I grew up with one. It's the only dog for me. I'll pay $10,000 for a well-bred one. But that's a lot of money. I have the money. I'm just not sure I'm ready."

Clearly, this man needed someone or something to care for. At that moment, it was his father. I hoped that a dog would be a bridge to purpose after his father's death. And as that dog grew older, hopefully he would find something or someone else, perhaps a grandchild.

At times, he broke down in sobs. Then he would hide his tears and take a long break between sessions, likely embarrassed at showing weakness.

After one of these outbursts, he said, "I think I'm done with this."

"We can reschedule," I said. "Or I can schedule you with another provider." I was near the end of my fellowship and would have to hand off active cases to my supervisor.

"No."

From the beginning, he had been concerned about the "merry-go-round" of providers at the VA. It was overwhelming for him to have to tell his story from the start, and this discouraged him from continuing.

"I've had enough," he said.

With that, he walked out.

To keep my own boundaries, I haven't checked to see if Clyde is still out there. I'm not sure that if he takes his life, I'll even be notified.

Despite these few failures, I felt that my WWS groups offered veterans who felt lost and alone the education, understanding, and hope for a better future.

CHAPTER 21

RUNNING WWS GROUPS

When I applied to the VA, a veteran supervisor had suggested that, in addition to providing group and individual therapy, I propose a process group to work out WWS concepts for other veterans.

Around this time, a close buddy, who had been unable to reconcile his need to redeploy, was getting divorced after twenty years of marriage. He was the fourth Marine buddy this year to get divorced. At this point, the knowledge I had accumulated about WWS had incubated long enough. It was time to put my money where my mouth was and start it up.

The first step was to begin writing this book. I did and quickly realized that I had more work to do developing my ideas. A year with the VA devoted to this process would help.

I sent a proposal to the VA.

WWS Proposal

I designed the group to provide psychoeducation about the Warrior Withdrawal Syndrome. In the proposal, I described the symptoms, the behaviors, and the attitudes of veterans with WWS who desired to return to combat largely because of the extreme challenges in adjusting to civilian life. Some of the WWS symptoms include the following:

- Fractured sense of self, identity, and purpose
- Anger management
- Communication challenges
- Maladaptive coping mechanisms to manage distress and strong emotions, like substance abuse and risky behavior

These symptoms matched the problems my supervisors at the VA were seeing. Joining the military was like joining a cult, leaving its former members with multiple challenges once they returned to civilian life. I would run the WWS groups with methods to deprogram from this quasi-cult, like an "exit interview" from the military.

Unlike other groups, mine would not be specifically aimed at providing coping skills or immediate relief but instead seek insight into the military training that led to the veteran's current level of functioning. At the same time, my groups might offer the veteran an opportunity to vent, providing a much-needed resource given the growing waiting lists for individual therapy at the VA.

I listed my proposed goals for the groups:

- **Describe symptoms.** Introduce WWS as the cluster of symptoms sometimes diagnosed as chronic adjustment disorder, PTSD, or other trauma- and stressor-related disorders.

- **Provide psychoeducation.** Describe how the process of indoctrination into the military culture had been entrenched in their psyches, making departure from the service and the transition to civilian life challenging.

- **Emphasize retraining.** Instill hope for veterans that they are not broken and therefore don't need to be "fixed." What they need is "retraining" from the military mindset that doesn't work in civilian life. By deprogramming the military mindset, some of the stigmas associated with mental health could be removed.

- **Foster insight.** Increase desire for change and help a pre-contemplative veteran (i.e., one who is not aware that a problem exists and who is not seriously thinking about changing or getting help) to gain insight and become contemplative (i.e., aware that there is a problem and directed toward overcoming it).

- **Address existential concerns.** Discuss issues like discounting the future, survivor's guilt, moral injury, and the fear of developing "invisible" wounds such as traumatic brain injury (TBI). Once in therapy, veterans would be able to talk through existential concerns and resolve questions such as "What was the point of it all?" and "What's the point of continuing to suffer?"

Number of Sessions

Initially, I organized the WWS group to be four group sessions, once per week for four consecutive weeks. With these four one-hour sessions, I had time to include some discussion among the veterans about the aspects of the syndrome, including how service felt like an addiction.

By attending these four sessions, veterans would demonstrate a willingness to work toward improving their mental health. This willingness would meet the minimum requirements for a referral to individual therapy with a counselor, psychologist, or psychiatrist if anyone wanted to continue this work.

Participants and Eligibility

The group was open to all veterans, with the exception of those with mental health challenges or cognitive impairment that prevented participation in the group setting or may cause excessive disturbances in session (e.g., threats of violence, excessive tangential speech persisting after redirection, and psychotic symptoms).

Although the goals were common to all, my targeted consumers, so to speak, were the veterans in the pre-contemplative stage who didn't feel they needed help and therefore weren't motivated to change. Hopefully, my group would enable them to see that the transition had been more difficult than they thought. Through this realization, they could enter the contemplative stage, where they became aware that they had a problem and could think seriously about overcoming it. At the same time, they would have not yet made a commitment to take action, because they weren't even aware of how to define the problem. At this point, the group would educate them on what was causing their anger, communication problems, and so on.

Veterans could opt out of the group requirement of four sessions if they insisted on going directly to individual providers. They would be placed on a waiting list and may receive a lower priority than those who completed the groups.

My Role

As group facilitator, I would be there less as a therapist than as a peer support to help them understand how their experiences in the military impact their current life negatively and figure out ways to adjust to civilian life better. As a peer support, I would share my own journey to provide insight into how military training had changed my thought processes.

Calls for Referrals

I posted referrals to the VA Long Beach health system that included some of the community-based outpatient clinics (CBOCs). Because I was able to offer the group virtually, I could accept Veterans from hundreds of miles away.

I sent out the following email to the mental health department distribution list:

> I am accepting referrals for an in-person co-ed Baseline Adjustment from Military Functioning (WWS) Group. WWS is a four-session, peer-support, psychoeducational group.

Unfortunately, given challenges with timing and diagnoses, I would only get the bare minimum to sign up. As the groups continued to progress, I went from one or two referrals to four or five. While the numbers seemed small, the group was picking up steam. Near the end of the fellowship, I had a short waiting list and ended up doubling the number of groups I offered per week to ensure that the information would get to the veterans before my fellowship was up.

Typical First Session

Attendees were on time to the second session or even early, most of them dressed in combat fatigues or cammies. Many of them swaggered into my group on the first day as if they owned the place. Those who wore civilian attire still had some identifier of their veteran status, whether it was a U.S. Navy baseball cap or a campaign ribbon sticker on their cane. After slouching down in chairs that were closest to the door, most of them crossed their arms, looked around for all exits, and measured the stature of everyone else in the room.

I made eye contact with each, nodded, and said, "Welcome."

After briefly introducing myself, I covered confidentiality, licensure, and other mandated points, and then I asked them to introduce themselves along with their branch of prior service and the deployments they had made.

To get conversation going, I would ask the group, "Anyone like to share why they're here?"

If no one responded, I called on someone. I always used their names; they all wore name tags. In addition, their names are displayed at the bottom of their photos in their online profiles.

Often, they would say something like, "I'm not sure why I'm here. I should be able to handle this shit on my own."

"I get it," I would reply. "Coming here makes you feel weak. That goes against all you've learned in the military."

"Yeah," they would all concur.

"So why are you here?" I would ask.

Many said things like "I don't understand what I'm going through. Except that I don't feel right." Or they would say, "My wife's going to divorce me if I don't get better control of my anger." Some admitted being sick and tired of feeling bad all the time.

I reassured them that they weren't crazy or broken. The just needed retraining.

To provide an umbrella under which they could find some cover, I would tell them my story. I started by listing my combat deployments and ascension from grunt to special operations, including my deeper association with JSOC.

Some were impressed.

"You were a snake eater, dude," said a thirty-eight-year-old pilot who had lost an eye during a skirmish in Iraq.

"I was," I replied. "I was trained by the best, and at the tip of the spear. Yet, I fell apart leaving active service. If I could, anyone could."

I impressed upon them that, though it was hard to talk about what was troubling them, it was far worse to have nightmares, startle at the sound of a doorbell, and have your wife leave you because you smashed in the bathroom mirror. "If you avoid seeking help for a broken finger, you will live with the pain and a poorly healed finger."

This opened something up, and the veterans would begin to spill out details about similar events, from destructive relationships to substance abuse.

How They Changed During Military Service

One of the first things I discussed with the group was how being in the military changed their personalities. We considered how it made them compliant and obedient, diminishing an autonomous sense of self from the moment they went to basic training or boot camp.

I helped each veteran see that a little veteran still lived inside their psyche. This homunculus had an affinity for service and wanted them to continue pursuing this obsession and rejoin service. They accommodated this internal voice by being a warrior in civilian life. For some, this might mean becoming law enforcement officers, a job that could hold the irrational desire to be a warrior at bay. For others, working out or wearing military gear might be a compulsion that reduces the distress of no longer being a warrior.

To demonstrate how the military hijacks a sense of self, I tell them to look down at their shoes. "How many of you still lace your boots left over right?" I ask them.

At least half of them laugh, because they're still doing what they were trained to do in boot camp. "There's no practical reason to do this," I explain. "But you're still doing it unconsciously to avoid that DI from yelling at you. How silly is that! There's no possibility of getting lambasted from him thirty years later, yet you do it because it seems to help. No one yells at you now for your boots."

I explain that it was an example of classic conditioning. "You look at your shoe. That unconsciously jiggles a memory of getting yelled at for lacing right over left. To avoid the punishment, you lace left over right. No one yells, so you repeat the behavior *ad infinitum*. I scream at my kids to *shut up*, even when I know there's a better way to stop their bad behavior. Yet I keep doing it, because when I yell, they shut up."

By this time, most would have let their guard down, and they would share stories of how they kept behaving as if they were in the military, from continued hypervigilance to aggressive behavior, even violence. They would talk about carrying additional equipment for the "just in case" scenarios, keeping their backs to the wall, and having heated interactions with kids, spouses, and coworkers. Some would express how, in retrospect, maybe they could've communicated differently.

"The key," I would tell them, "is to untrain automatic, gut behavior that no longer works in civilian life. The tying of shoelaces is unimportant. But hypervigilance and aggressive or even violent behavior are not. Yelling at your buddies may have been normal military communication, but yelling at your coworker, your kids, or your spouse is now considered aggressive."

"Yeah, but I get so pissed off," said "Dan," a jumpy forty-eight-year-old bartender, too old to return to service, who has been chewing gum nonstop. "This motherfucking waiter always comes in late, usually because he's been on a bender. And I almost got fired for lambasting him in front of customers. I can't help it. I'm never late one second. It shows disrespect."

Nods all around.

"Yep," I agreed. "Being on time was ingrained into our psyche. Being late could mean the difference between life and death during deployment. Small mistakes in how the Iraqis prepared their munitions meant that I was still alive. Had they aligned their artillery cannons slightly more accurately or selected the high-explosive variable time rounds for troops in the open, I wouldn't have made it. Later, as a fire support coordinator, I'd see how helicopters could be taken down with our own indirect fires if they were not on our timeline."

"Yeah," said another vet, anxiously shaking his leg. "I flew a helo. Always got to my bird early."

Others pitched in with stories about never being late.

"The thing that riles me big time," said one ex-Navy vet working in law enforcement, "is how some guys at the station goof off. A call will come in about a murder, and these guys will keep jabbering about how they got pussy. The sergeant often has to yell at them to get their asses moving. They're not mission-focused."

"Copy that," said a young sailor who worked in construction. "I hate guys who cheese-dick the job. Takes them an hour to hit a few nails. Makes me want to eliminate them."

"Does that murderous desire make you feel crazy?" I asked.

Most raised hands. A few shared how rage made it hard for them to keep a job, leading to them quitting or being fired, or how the police had been called to their homes to manage intense arguments between them and their spouses. Virtually everyone described times when they had a plan to "take someone out."

"Let me share my own story about having murderous tendencies," I tell them. "While on active duty, I sought out Craig Bryson, my unit psychologist,

216

because I was having homicidal thoughts about some of the Marines who were preventing me from achieving my objectives with JSOC."

I described how I had laid out the "mission planning" and was wrestling with a few moral objections. Mission planning for my JSOC unit meant coming up with the concept of operation for a clandestine raid that would result in the kill or capture of a terrorist. I was now applying that same level of planning, going through mission stages—"find, fix, finish, exploit, assess, disseminate," or F3EAD—to overcome obstacles in my JSOC path.

"Are you going to do it?" asked Craig.

"Of course not," I said.

"Then you're fine. You're just executing the planning that we taught you do to over the past few years."

I sighed with relief. I wasn't "crazy." Having homicidal thoughts must be normal. I'd gone through training for close target reconnaissance, learned skills to make clandestine entries into buildings, worked with units that found the most effective ways to dispose of bodies, and practiced these skills on U.S. soil to use overseas. My default reaction to someone pissing me off was figuring out where they lived and eliminating them.

"That can't be good," I said.

"Nope," they replied. They slowly but eventually shared their own stories about coming close to murdering someone. One described a bar fight where he went ballistic and punched a guy unconscious. He couldn't stop until other guys broke up the fight. Another described slamming into the car in front of him because the Arab driver wearing a head scarf had been on his cell phone at the light, and he didn't move when the light turned green.

What It Means to No Longer Wear Their Uniform

I brought up how confusing transitioning out of service could be. "You're likely asking yourself questions like 'Am I still a warrior?' and 'How can I achieve the warrior recognition I used to have?' One solution is to still wear the uniform."

They laughed, given that several wore fatigues or cammies to the session.

"But you need to exercise some caution. Chest candy [or medals] on your lapel, a baseball cap with an American flag, and a T-shirt with guys in cammies are okay. T-shirts with skulls on them or a 'come take them' shirt with a rifle on it will broadcast your warrior status, but they might also make people shun you as some kind of monster."

Some looked at their T-shirts and chuckled. Several wore shirts with the Gadsden flag on them; others had "grunt style" shirts on. One even had a shirt on that said "Back Off" with two middle fingers underneath it. Even their tattoos were challenges to others and subtle hints to treat them as monsters. "Sheepdog" or "USMC" tattoos were in highly visible places. Others had tattoos of blood and skulls all the way down their forearms to keep people away. One veteran had a tattoo of a Glock pistol on his forearm and made a joke of having to "clear his weapon" before coming into the sessions.

"So," I asked, "how can you keep your warrior identity while fitting into civilian life?"

One veteran discussed how he went to a nonprofit that would pay for tattoo removal. He realized that he wanted to climb the corporate ladder, and having two full-sleeve tattoos in that environment was unacceptable. Others had the opposite reaction, saying how disrespectful it was that their jobs required they cover up their tattoos with long sleeves or a bandage, as if they had to "cover up" their service.

I asked one veteran what his objective was with his "Back Off" T-shirt.

"I want people to stay the fuck away from me," he answered.

"Because . . ."

"I don't know. They piss me off."

"Who do you want to keep at bay?"

"Not other veterans."

"What about your family?"

"No, I don't want them to back off, either."

He was stuck in that same catch-22 that I was in. Even though he wanted close relationships, being alone was easier, and he was embracing the "monster" personality to achieve it.

What Is Permanent and What Can Be Modified

"Okay," I said. "Let's talk about which behaviors you wish to keep and which you want to change."

Most concurred that they wanted to keep a good work ethic, punctuality, loyalty to peers, having another person's back, and so on. They wanted to change their hypervigilance, quickness to anger, challenges in communication, and alcohol addiction, among other destructive habits.

"What about your language?" I asked. "Can you replace 'motherfucker' with 'bad dude,' or something like that?"

They laughed. "No fucking way. Not giving that up."

I laughed, too. "It's hard to change habits. To avoid using 'motherfucker' as every other word in our vocabulary. To not scan the room and rooftops to feel safe. To not shout when communicating. To not slam the door when our wife annoys us. To not turn to the bottle for comfort."

Then I asked, "Anyone want to share what current habits they have that they want to change?"

Some talked about drinking too much, getting into too many barroom brawls, not going to the gym to stay in shape, or always preparing for the worst-case scenario.

"My rifle is never far away," said a forty-something soldier, chain-smoking nonstop.

"It's programming," I said, "that needs to be retrained. While we had to stuff grenades in our pockets when we left the wire, there's no need to carry around an M-4 rifle when meeting your buddies at the neighborhood bar. At the same time, some things might never change. We might always seek the seat closest to the door in the restaurant and scan the room for any evil-looking dudes."

"That's good, isn't it?" asked a thirty-five-year-old Green Beret. "We'll always be the first responders should any bad guys try to pull a fast one."

They all responded, "Copy that."

I gave them homework to ponder further and write down what habits and behaviors they wished to change.

Combat Addiction

At some point, I would bring up the problem of combat addiction. I would start by again talking about the homunculus that still lives inside their psyches and how that little veteran wants them to continue to pursue this obsession and rejoin service.

In civilian life, the little warrior might become a law enforcement officer, a profession that partly satisfies the irrational desire to be a warrior. For others, working out or wearing military gear might be a compulsion that reduces the distress of no longer being a warrior. Still, in spite of these accommodations, the little veteran's voice often continues to shout "*war, war, war.*"

"How much thought do you give to redeploying?" I asked them.

A forty-two-year-old Marine, still sporting the standard jarhead, "high and tight" buzz cut, quickly jumped in. "I'm looking for ways to increase my odds that my number will be called when the president goes back to the individual ready reserve to select troops for the next operation in Ukraine or wherever else."

Several others nodded. One muttered, "Roger that."

I mentioned how common that is and how civilians find it impossible to understand. Only those who live it understand it. War makes us feel most alive.

"Yeah!" they agreed.

I relayed the story about the first time I met someone obsessed with war. Mike Jones, a sergeant major from JSOC, came to talk to us about PTSD while I was still on active duty.

I knew Jones by reputation. He had been drummed out of the unit for operational security violations. Ultimately, his struggles with PTSD led to some near-catastrophic misses, costing him his security clearance and position.

He spoke to us about the continuous deployments, the unrelenting cycle of being deployed with JSOC, returning home and having only a little time to rest and recover before transitioning back to training for the next deployment. At times, the turnaround for JSOC was six months between deployments. And of those six months "at home," many were spent on training evolutions.

He did not talk about any one particular traumatic instance, no raid or event that he highlighted as the reason for PTSD. Nevertheless, the experts—the VA and psychologists—had labeled his anger-management issues and substance abuse as comorbid with PTSD.

"Roger that," said one young Marine.

Mike volunteered for more deployments, many into combat, because he experienced less stress while overseas than at home. Though being overseas and down range was life or death, the mission was all that mattered. His mortgage, kids' orthodontia, and spouse's needs didn't come up. He even tried to ensure that he would be deployed during the holiday season so he wouldn't have to be around his family.

In focusing only on the mission, he could find clarity of purpose and get through the day-to-day suffering with the knowledge that it was meaningful. When he got home, his purpose was not immediately apparent, and the anger, doubt, and trauma would creep in. He tried to hold them at bay with alcohol. His drinking led to being late to work and performing less and less efficiently. Still, he would go on another deployment.

Ultimately, he made an error of judgment. While deployed, he kept a mistress overseas, thinking she would never come to the United States. With his mind clouded, he thought it would be fine to tell her exactly where he lived in the U.S. He'd incorrectly reasoned that his mistress would never look him up. She did. That led to a heated confrontation between her and his wife, and his military career came to an end.

Mike came to talk about how PTSD looks different for different people. He realized that he wasn't recovering from his deployments.

"Today," I said, "I would diagnose him with WWS, combined with a PTSD diagnosis. How many of you have a similar story?"

Several hands shot up.

"Big difference now, though," I continued. "No one was talking about WWS at that time, and Mike didn't know how to ask for help. Luckily, you all have this group to learn what makes you have this itch and how to get help."

At this point, we got into a discussion on the split between wanting the comforts of home and craving the thrill of war. Some of them would count the days until they were deployed, then while on deployment, count the days until they were home. As soon as they got home, they'd start counting down to the next deployment. It couldn't come fast enough.

Typical Second Session

By the second group session, their posture would change almost as soon as the conversation got going. Some leaned forward into the discussion, others became animated and forgot the time because they had so much to say.

I'd go through the eight components of relationships with intimate partner violence (IPV): control, dependence, digital monitoring, dishonesty, disrespect, hostility, harassment, and intimidation. I drew an analogy to the careers of the veterans as the starting point for discussing how their time on active duty was destructive.

"Let's talk about the military relationship, about how it's as if you guys were behaving like you were in an intimate relationship with a partner that kept beating you, insulting you, humiliating you."

This comparison sparked a lengthy discussion about commander and DI abuse, as well as how when they were on active duty, they felt like an abused spouse, married to the service.

"For most people, 'just leave' is the simple answer to how to respond to abuse. 'Just get out,' their spouse, family, or others say. But many don't feel like they can, or they worry about their safety if they do. This is similar for veterans. Something else keeps you on active duty. Something allows you to put up with control, disrespect, hostility, etc. It's the same as substance abuse, though the substance isn't cocaine but warrior addiction."

Nods all around. I then went through the eleven components of substance abuse. Most experienced all eleven, with significant withdrawal and tolerance symptoms, indicating severe addiction to being a warrior. At the same time, many *were* addicted to substances, particularly alcohol, so it was a double whammy.

"Hey, guys," I said, "I know many of you have an alcohol problem. But in thinking about your addiction to war, what might the substance underlying warrior addiction be? For me, for instance, my substance was the need to be validated for being tough. I'm embarrassed to tell you that my time in my last marathon was just under six hours. Though I had knee surgery only a month and a half before running in the race, I've been used to coming out on top. I wanted to return to war because I felt I hadn't been in the thick of it and was no hero."

Some would cite "valor" or "honor" as their substance. Several recognized that their combat deployments as contractors didn't provide the same feedback as being a Marine, sailor, or soldier had—and offered no "medal."

"I need to feel rewarded," said one half-bald and paunchy fifty-year-old ex-Army soldier. "So I kept reenlisting, hoping to redeploy, even though I'd get paid less than a quarter of what I was earning as a contractor."

"Copy that," said another.

Some stated "pride" or "commitment" as their substances.

One Vietnam veteran said that he just missed the simplicity. "Everything made sense over there. I was either on patrol or in the rear. I didn't get to choose what I ate. I had two pairs of socks to alternate between. It was just easier." Being a soldier absolved him of all anxiety related to choice. Even though he could be killed by the Vietcong at any point, he'd trade the last fifty years of "safety" living in the U.S. to go back. He had no ability to recapture that feeling. Where could he go? What could he do that would be the same?

Virtual Therapy

In addition to in-person groups, I conducted virtual groups. We sat at our computers, using the VA's proprietary video connection system.

Most of the time, I conducted sessions from a computer in my office at the VA. My office was repurposed from what had previously been a "hotel at the hospital," or "hoptel," for veterans who made longer trips for their appointments and stayed overnight. Given the building's former use, the nurse's stations had once been a laundry room, another therapist was working out of a broom closet, and the rest of the offices were previously one-bedroom rooms that shared a bathroom with their adjoining room.

My office happened to be next door to the ketamine clinic. At any time during the group sessions or with individual patients, the loud sound of a toilet flushing was a probability. Although not too embarrassing for a Marine—after all, the "toilets" we'd used in Iraq had no accommodations for modesty—I was embarrassed now as a mental health professional. Someone could be in the middle of explaining their trauma when—*whoosh*—the toilet would flush.

After allowing a few minutes for all the veterans to log in, I would start. For the most part, I would stick to the material I had. When I could, I shared my screen to show them what I was talking about. Without a co-facilitator who could keep an eye on the veterans' engagement while I was presenting the material, I couldn't tell how they reacted. Still, given their engagement with the topics and the personal stories they shared, they seemed to be paying attention.

In one group, I ended up with three Marine veterans and one sailor. Their ages varied from twenty to fifty-eight. Of the four, only one had seen combat. Their experiences varied from the military of the 1990s to active-duty service that the youngest veteran had just completed the year before.

Two had not married and didn't have any kids. The older one reflected on how he'd put his career as a civilian ahead of trying to establish a family and had focused on being the best salesman he could be. By focusing on this goal, he postponed his WWS need to redeploy. He traded the warrior "substance" for being a "workaholic," and he succeeded. It was an easy bridge for him, as he'd been successful as a Marine and kept up that "110-percent effort" into retirement.

When I conducted the triage with him before he joined the group, he'd brought his elderly mother with him. While he was careful not to use foul language around her, once we started trading military terms and acronyms, his language became less polite and controlled. At one point, he snapped at his mother, saying, "You don't get it. He does."

Recognizing that WWS may have cost him a family of his own and damaged his relationship with his mother, he signed up for the next WWS group to focus on some of those issues. He was an ideal candidate for individual therapy to process the lost opportunity of a family. He could set interpersonal effectiveness as a short-term goal and learn how to use assertive communications with his mother when he wanted to ask for or decline something.

Protecting My WWS Tendencies

The youngest Marine in the group was in the middle of working through his VA disability claims. He asked to stay online longer with me to talk through some of the choices he was making regarding that process.

While I was happy that I'd earned his trust, this wasn't my field of expertise, and I couldn't keep the next appointment waiting. I did, however, take extra time to direct him to a VSO that might be better able to answer his questions about claims.

Though it was hard not to take the extra time to help him, I needed to protect myself from my own WWS obsession.

When I worked for the county, it was easy to call it a day at 5:00 p.m. As soon as I started working with veterans at the VA, though, the addiction started to flare again. Unable to set appropriate boundaries, I came home bearing their problems, worrying about how to help them out of their post-service WWS hole. Bending over backward to accommodate the veterans, I pushed the limits on the length of sessions, called more frequently than required, and so on.

Understanding why I felt compelled to act this way was my first step toward handling my addiction. By saving other vets from WWS, I believed I would heal myself. At the same time, I knew this mindset was dangerous for my own mental health. I needed to build a boundary around these cases, or I would march quickly to burnout, something veterans working at the VA commonly experienced when they went above and beyond their schedules to assist veterans. Protecting and assisting veterans was a way to get that warrior fix, and it came with additional withdrawals.

It was particularly hard to fight the urge to check on veterans who had discussed suicidal ideations with me and then terminated care. I knew their names and could google search to see if they'd taken their own lives. I refrained because following up on them would plunge me back down the warrior withdrawal rabbit hole—taking me away from working with my son on a project or helping my daughters set up their Barbie dolls.

Last Sessions

By the last couple of sessions, many had opened up and expressed regrets: going to combat; not going to combat; joining the military; leaving the military and letting their buddies down.

I talked about how the "coulda, woulda, shoulda" leaves us stuck in the past. Our goal was to try to design a better future for ourselves.

"Pain is inevitable, life is painful, and we choose to suffer," I said. "Dragging the pain of deployments and the past into the present is an avoidable suffering."

This reframing would set off a discussion of steps they could take, however small, to make their lives more tolerable and even pleasurable.

Most of the veterans had completed their homework of making a list of the behaviors they were interested in changing, and we discussed how to turn those into short-term goals with their individual providers. Sometimes it took a while to get from "I want to be less angry" to "I am interested in learning adaptive coping skills for managing strong emotions"—or from "I don't want to punch holes in the wall anymore" to "I struggle with distress tolerance and want to improve my ability to manage stress in a constructive manner."

Every veteran left with a referral to individual therapy. I heard back from some of their therapists regarding how helpful the group had been. One provider told me about a graduate of the group who was progressing well, and she said she'd talked to him about attending a coping-skills group.

"No thanks," he told her, "I don't do groups."

"But you did the WWS group?" she asked.

"That's different," he replied.

I smiled with satisfaction.

Another testament to my success: by the end of the last session, most of the veterans were supporting each other and finding ways to connect outside of the group.

EPILOGUE

I began my WWS groups in January 2023. At the time of this writing, I've conducted more than ten of them. Graduates of the program have demonstrated better insight into their lives as well as improvements in anxiety and depression. PTSD symptoms have reduced to near zero on self-report measures such as the Generalized Anxiety Disorder-7 scale, Patient Health Questionnaire-9, and PTSD Checklist for DSM-5. While not all veterans responded to requests for completing self-reports, every veteran who did so noted a marked change and decrease in symptoms, particularly on the checklist.

Despite the successes, a large population of WWS veterans is out there struggling with little help or understanding of their issues. I hope this book will help them to recognize that they're not alone or broken, and that they don't need to be angry all the time. This understanding will hopefully serve as an impetus to seeking treatment.

I'll reach some through family members who read the book and level with them, much like my ex-wife did when I said I was a peaceful duck on top of the water and she said, "You're an angry duck on top of the water." Spouses will hand the book to their veteran partners and say, "This is you. Now get help to fix yourself."

While working at the VA with veterans is rewarding, my larger goal is to get the WWS message to active-duty service members before they leave the service. In this way, I can provide the ounce of prevention that led me to initially want to be a child psychologist.

I don't know if I'll ever fully recover from my itch to return to combat. Even today, having to depart from active duty feels like an inglorious end, a career cut short. But I have learned how to better control my feelings. The more groups I've run, the easier it has gotten to set boundaries, end sessions on time, manage the discussions, and refrain from being participants' single point of contact for all things VA.

Luckily, I have my profession as a method to escape and my kids to spend time with. For every veteran I've talked to, I'm reminded that there are likely just as many spouses in desperate need of better understanding their partners.

ACKNOWLEDGMENTS

T hroughout life, I have been aided by countless individuals, too many to list individually. From leaders while I was on active duty to the supervisors in the psychological community who helped me learn a new profession, all pushed me to this point where I'm finally able and willing to tell this story. My family and friends have supported me throughout. Without them, none of this would have been possible.

NOTES

1. Abraham Maslow, *Motivation and Personality* (Delhi, India: Pearson Education, 1987), 64.

2. David Kendrick, "Why Veterans Feel Addicted to Combat" *Psychology Today* retrieved from: https://www.psychologytoday.com/us/blog/the-veterans-corner/202201/why-veterans-feel-addicted-combat.

3. Matthew Bowen and Angelica Chang, "Combat Addiction: Revisited and Reaffirmed," *Journal of Psychology and Clinical Psychiatry* 7, no. 6 (2017): 3.

4. Lionel Solursh, "Combat Addiction Post-Traumatic Stress Disorder Re-Explored," *Psychiatric Journal of the University of Ottawa* 13, no. 1 (1988): 18.

5. William Killgore, et al., "Post-Combat Invincibility: Violent Combat Experiences Are Associated with Increased Risk-Taking Propensity Following Deployment" *Journal of Psychiatric Research* 42, no. 13 (2008): 1119, doi: 10.1016/j.jpsychires.

6. William Killgore, et al., "Preliminary Normative Data for the Evaluation of Risks Scale—Bubble Sheet Version (EVAR-B) for Large-Scale Surveys of Returning Combat Veterans," *Military Medicine* 175, no. 10 (2010): 730, doi: 10.7205/milmed-d-09-00143.

7. Morten Brænder, "Adrenalin Junkies: Why Soldiers Return from War Wanting More," *Armed Forces & Society* 42, no. 1 (2016): 21.

8. Bessel van der Kolk, et al., "Inescapable Shock, Neurotransmitters, and Addiction to Trauma: Toward a Psychobiology of Post Traumatic Stress" *Biological Psychiatry* 20, no. 3 (1985): 319, doi: 10.1016/0006-3223(85)90061-7.

9. Morten Brænder, "Adrenalin Junkies: Why Soldiers Return from War Wanting More," 7.

10. Matthew Bowen and Angelica Chang, "Combat Addiction: Revisited and Reaffirmed," 2.

11. Amy Adler, et al., "Effect of Transition Home from Combat on Risk-Taking and Health-Related Behaviors" *Journal of Traumatic Stress* 24, no. 4 (2011): 387, doi: 10.1002/jts.20665.

12. Austin Jenkins, "Risk-Taking Common Among Returning Soldiers," *NPR* online, August 4, 2011, https://www.npr.org/templates/story/story.php?storyId=130075266.

13. Maria Steenkamp, "True Evidence-Based Care for Posttraumatic Stress Disorder in Military Personnel and Veterans," *JAMA Psychiatry* 73, no. 5 (2016): 431-2, doi:10.1001/jamapsychiatry.2015.2879.

14. *The Myth of Sisyphus and Other Essays*, New York: Alred A. Knopf, 1955 p.3